Trends in Logic

Volume 37

For further volumes:
http://www.springer.com/series/6645

TRENDS IN LOGIC
Studia Logica Library

VOLUME 37

SCOPE OF THE SERIES

The book series *Trends in Logic* covers essentially the same areas as the journal *Studia Logica*, that is, contemporary formal logic and its applications and relations to other disciplines. The series aims at publishing monographs and thematically coherent volumes dealing with important developments in logic and presenting significant contributions to logical research.

The series is open to contributions devoted to topics ranging from algebraic logic, model theory, proof theory, philosophical logic, non-classical logic, and logic in computer science to mathematical linguistics and formal epistemology. However, this list is not exhaustive; moreover, the range of applications, comparisons and sources of inspiration is open and evolves over time.

Rafal Urbaniak

Leśniewski's Systems of Logic and Foundations of Mathematics

 Springer

Rafal Urbaniak
Centre for Logic and Philosophy of Science
Ghent University
Ghent
Belgium

and

Department of Philosophy,
Sociology and Journalism
Gdańsk University
Gdańsk
Poland

ISSN 1572-6126   ISSN 2212-7313   (electronic)
ISBN 978-3-319-34416-4   ISBN 978-3-319-00482-2   (eBook)
DOI 10.1007/978-3-319-00482-2
Springer Cham Heidelberg New York Dordrecht London

Printed on acid-free paper

Springer is part of Springer Science+Business Media (www.springer.com)

To my teachers

Preface

The Lvov-Warsaw School of Logic and Analytic Philosophy was one of the most important schools of philosophical thought in twentieth century. In early 1910s its members already discussed the validity of the principles of excluded middle and contradiction. Among ideas developed in this school, one might count Łukasiewicz's view that one can believe a contradiction and that certain sentences can be neither true nor false. This led to the construction of his three-valued logic. Another example is Ajdukiewicz's conventionalism about meaning and his formal work on definitions (it seems that it was Ajdukiewicz and Łukasiewicz who first focused on the consistency, translatability, and non-creativity conditions on definitions, at least on the Polish ground). Other examples include Jaśkowski's approach to natural deduction and his work on discussive logics, Lindenbaum's lemma on maximally consistent sets of formulas, Presburger's work on arithmetic, Kotarbiński's semantical reism, and Tarski's work on formal semantics and truth.

One of the representatives of this school was Stanisław Leśniewski (1886–1939) (Alfred Tarski, whose importance in twentieth century logic it is hard to overestimate, was his only Ph.D. student). Leśniewski developed his system of foundations of mathematics as an alternative to the system of *Principia Mathematica*. He constructed three systems: Protothetic, which is his version of a generalized propositional calculus, his own (higher-order) logic of predication called Ontology, and a theory of parthood called Mereology.

Leśniewski's work is interesting for a few reasons.

- If one is interested in history of logic in general, it is hard to deny that Leśniewski was one of the key figures in one of the most important schools of logic in twentieth century. He devoted his research to developing an alternative to the system of *Principia Mathematica* and this attempt is worth studying in his own right.
- If one is interested in the development of Tarski's thought it might be useful to learn what his Ph.D. supervisor's views were and how Leśniewski's work and Tarski's ideas are (or are not) related.
- Philosophical discussions in which Leśniewski participated pertained to issues which are discussed quite lively even today. His approach to semantical and set-theoretic paradoxes and his views on the validity of the principle of excluded middle and of the principle of contradiction are philosophically interesting.

- Leśniewski was a nominalist and his systems were a nominalistic attempt to provide a system of foundations of mathematics. It is a major attempt of this sort and as such it is worth an examination.
- His metalogic is quite specific. Nominalist as he was, he wanted to develop a purely inscriptional syntactic description of his systems in a way that did not make any reference to expression types. It is interesting to see how he proceeded.
- His systems have some interesting properties. For instance, in all of them definitions can be creative (and this is not considered to be a problem). The generality of Prothetic admits interesting extensions (intuitionistic (see López-Escobar and Miraglia 2002) or modal (see the works of Suszko and in general, see Sect. 3.7 for references). The language of Ontology (which, in a way, can be viewed as one of the first free formal logics) is, arguably, more suitable for capturing certain aspects of predication and abstract noun phrases as they work in natural language.

This book is devoted to a presentation of Leśniewski's achievements and their critical evaluation. I discuss his philosophical views, describe his systems, and evaluate the role they can play in the foundations of mathematics. It was my purpose to focus on primary sources and present Leśniewski's own views and results rather than those present in secondary literature. For this reason, later developments are not treated in detail but rather either mentioned in passing, or described in sections devoted to secondary literature included in some chapters. The intended audience of this book includes philosophy majors, graduate students, and professional philosophers interested in logic, mathematics, and their philosophy and history.

Parts of this book started as my Ph.D. dissertation written under the supervision of Richard Zach and defended in 2008 at the University of Calgary. Other parts report on research which went beyond the dissertation (in particular, Chap. 6 was written together with Severi K. Hämäri). Ultimately, in 2011 my wife took me to India, where she pursued her research in Indian philosophy and forced me to use those few months to write the whole book anew.

Varanasi, India, March 2012   Rafal Urbaniak
                                                        Department of Philosophy
                                                       Sociology and Journalism
                                                            Gdańsk University
                                                             Gdańsk, Poland

                                        Centre for Logic and Philosophy of Science
                                                             Ghent University
                                                             Ghent, Belgium

References

López-Escobar, E. & Miraglia, F. (2002). Definitions: The primitive concept of logics or the Leśniewski-Tarski legacy. *Dissertationes Mathematicae* 401, Polska Akademia Nauk, Instytut Matematyczny, Warszawa.

Acknowledgments

This book started as a Ph.D. thesis written under supervision of Richard Zach at the University of Calgary and I owe him gratitude for his time, effort, and patience. I am also grateful for all the comments on my dissertation which I received from Nicole Wyatt, Jack MacIntosh, and John Kearns.

I would also like to thank my other teachers. Words cannot express my gratitude to my first philosophy teachers, Martyna, and Robert Koszkało, from whom I stole (well, borrowed and never returned) my first book on Leśniewski's systems and without whose motivation I would probably end up being a helicopter pilot (and we all know how unexciting and underpaid that job would be). I owe gratitude to my M.A. supervisor, Jarosław Mrozek, who allowed me to teach my first seminar on Leśniewski before I graduated and to my M.A. referee, Andrzej Włodzimierz Mostowski.

As for Chap. 6, many thanks for their comments to John MacFarlane, Nuel Belnap, Wilfrid Hodges, Paolo Mancosu, Øystein Linnebo, and Jan von Plato. Work on this part was supported by the Special Research Fund of Ghent University through project [BOF07/GOA/019].

Later parts were discussed at various places where I was giving talks: Ghent, Riga, Edinburgh, Szklarska Poręba, Hejnice, Geneva, Frankfurt, Dublin, St. Andrews, Auckland, and Oxford, among others. I am grateful to all the audience members for their feedback. The approach to higher-order quantification discussed in the last chapter was also partially shaped in discussion with Øystein Linnebo, who was my host at Bristol during my British Academy Visiting Fellowship, Hannes Leitgeb and Leon Horsten—other Bristol Philosophy Department members at the time of my stay in Bristol—and Stewart Shapiro, whom I ran into and tended to disagree with on various occasions.

I am grateful to my students, Paweł Pawłowski and Paweł Siniło, who proofread major parts of this text and helped me to avoid at least some of the mistakes.

I was also able to discuss a late draft of this book with Peter Simons during my stay in Dublin as a Long Room Hub Visiting Fellow, so I would also like to thank the Long Room Hub and Peter Simons for their support.

I am deeply thankful to my parents, Halina, and Andrzej Urbaniak—the list of things I am grateful for would probably be longer than the book itself. Finally, I am grateful to my wife, Agnieszka Rostalska, for her patience in general, for causing my voluntary exile in India and for turning it into a writing spree.

Contents

Chapter 1
Introduction

Abstract I provide motivations for the research, describe the structure of the book, give a short biography of Leśniewski and provide details of primary sources.

1.1 A Few Words of Motivation

Of course, other surveys of Leśniewski's work exist. However, to the best of my knowledge, their scope and interests differ from mine. Most of the discussions present in the literature focus just on selected aspects of Leśniewski's work. For instance, Słupecki (1953, 1955), Iwanuś (1973), Rickey (1976) and Lejewski (1958) focus on one of the logical systems only. Luschei (1962) and Betti (2005) cover not only the content of what Leśniewski actually published, but also almost indistinguishably speak of what he is known (from other sources) by the authors to have said or believed. Woleński (1985), Rickey (1976) and Betti (2005) are too brief to provide a complete introduction. When the above authors discuss Leśniewski's views, they focus on presenting them rather uncritically, whereas works that criticize Leśniewski (such as Grzegorczyk 1955) have been justly accused of misrepresenting the systems. So, my intention is to provide a survey that covers many aspects of what Leśniewski published, makes a clear istinction between what views can be assigned to him based on textual evidence and what he is reported or believed to have claimed, and, when is philosophical views are discussed, assesses them critically.

There are a few books on certain aspects of Leśniewski's thought written in French: Miéville (1984, 2001, 2004, 2009), Miéville and Vernant (1995), Gessler (2005, 2007), Peeters (2005). Their focus, however, is more historical than critical (see e.g. a review by Quinon (2011)), while my interests are more analytical.

Probably the only published book-long treatment of similar scope written in English is Luschei's book, titled *The Logical Systems of Leśniewski* (1962). In many respects it is admirable. Unfortunately, it has also been criticized on a few points.

R. Urbaniak, *Leśniewski's Systems of Logic and Foundations of Mathematics*,
Trends in Logic 37, DOI: 10.1007/978-3-319-00482-2_1,
© Springer International Publishing Switzerland 2014

First, the text has been found to be too polemical towards other logicians Cohen (1965).[1] It also does not contain an accessible account of how proofs in Leśniewski's systems work Thomas (1967), Kearns (1973) points out that not too many results are clearly stated.[2] Dawson (1965) emphasizes that the author gets involved in a somewhat biased philosophical commentary about the system and historical remarks rather than presenting Leśniewski's views critically.[3] I do not want to suggest that I managed to avoid any criticism of the above sort. Rather, I would be glad if it turned out that at least to some degree my work constitutes even a slight improvement on these points.

As I mentioned in the preface, the intended audience of this book includes philosophy majors, graduate students and professional philosophers interested in logic, mathematics and their philosophy and history. This puts the author in a slightly difficult position. On the one hand, the nature of the topic may make the book sometimes challenging for a philosophically-minded reader. On the other hand, due to accessibility considerations, mathematically-minded readers mind find the book slow-paced and sometimes not as technically precise as they think it should be.

As a philosopher, I am mostly interested in those technical issues which I think are philosophically relevant. Given that the intended audience includes philosophers, I decided to abandon some details of certain technically interesting dimensions of Leśniewski's work (like a step-by-step explication of his Terminological Explanations, detailed proofs of some theorems, or the technicalities related to Leśniewski's approach to formal systems as developing in time). This, to some extent, is a matter of taste—one person's *sloppy* is another person's *saving 15 minutes of my life*.

A related worry is that readers with purely mathematical background might consider philosophical aspects of this work to be of questionable relevance and validity. But this seems to be the attitude they would have towards most of philosophical

---

[1] "In his zeal to demonstrate the outstanding merit of Leśniewski's contributions to logic...Luschei sometimes devotes a disproportionate amount of space to denigrating the work of other logicians."

[2] "The book is expository. It does not contain proofs, and not many results are stated precisely. The author's style makes reading difficult; many passages give the impression of being literal translations from some language unlike English. The book is also marred by being unduly polemical. It would have been relatively simple to present Leśniewski's views and describe his practices, and then show how these differ from those of other logicians. Instead Luschei champions Leśniewski's views, and mentions many other logicians only to show how far short they fall of a Leśniewskian ideal...Luschei's explanations are less helpful, than they might be, for he uses a cumbersome, unfamiliar terminology, which sometimes makes his explanation as much trouble to work through as the T.E [Leśniewski's *Terminological Explanations*]. Luschei also fails to provide helpful comparisons with more familiar systems."

[3] "From its title, one might be prepared to find in this book either a presentation and development of the systems in question, or a discussion and commentary about the systems (or both of these). In fact, Professor Luschei concerns himself almost entirely with the latter task, and indeed no proofs of theorems appear in the text at all, although some examples of 'Leśniewski's technique of "natural deduction" ' are given in a note. Thus the book is primarily directed at readers who wish to learn something of the historical and philosophical background and foundations of Lesniewski's systems of logic, rather than to those who wish to become familiar with specific theses of the logic or to acquire facility with the formal techniques involved."

texts and I can hardly do anything about it. Just as some mathematicians would find philosophical discussion irrelevant and uninteresting, some philosophers would take highly abstract and philosophically uninterpreted mathematical results to be uninteresting and irrelevant. Having said this, I hope philosophically-minded mathematicians will also be able to find something interesting in this work.

1.2 The Structure of This Book

This chapter continues with a brief biography, a survey of primary sources, and acknowledgments. In Chap. 2, I discuss Leśniewski's philosophical views from the early period. I start with an account of Leśniewski's understanding of linguistic conventions (Sect. 2.2). Leśniewski's linguistic conventions may be thought of as various additional restrictions put on natural language. Those include definitions, which determine how he uses some terms, some general conditions on what truth conditions of various statements are, and some other assumptions about the language of the discourse which are hard to classify. I also provide some basic conventions that he accepted and used in the years 1911–1914. Next, after a brief detour through Leśniewski's views on proper names, I show how he applied those conventions to various problems. That is, I discuss what conclusions those conventions helped him to draw about existential propositions (Sect. 2.4), the principle of contradiction (Sect. 2.5), the principle of excluded middle (Sect. 2.6), the eternity of truth (Sect. 2.7), and how he attempted to solve a few paradoxes (Sect. 2.8): Nelson–Grelling's, Meinong's, Epimenides' (the Liar), and Russell's. Finally I say a few words about his rejection of abstract objects (Sect. 2.9).

Chapter 3 is devoted to Leśniewski's generalized system of propositional calculus (called Protothetic). In Sect. 3.1, I describe the historical context in which it was constructed. In Sect. 3.2, I discuss the notion of semantic categories which is crucial for the construction of Leśniewski's systems. Section 3.3 deals with Leśniewski's original "wheel-and-spoke" notation. Section 3.4 describes Leśniewski's motivations for his formulation of Protothetic (especially his criticism of *Principia Mathematica*). Protothetic originated as the result of gradual axiomatic generalization of the propositional calculus. The first propositional system ($\mathfrak{S}$) given by Leśniewski is a purely equivalential propositional calculus without quantifiers. The second system ($\mathfrak{S}_1$) admitted quantifiers binding propositional quantifiers and also variables representing propositional connectives. Extending $\mathfrak{S}_1$ with a certain rule (rule η) results in a stronger system $\mathfrak{S}_2$. However, η was quite complicated, so it was replaced by another, simpler rule (called the rule of extensionality, $\eta\star$). The result of this replacement is called $\mathfrak{S}_3$. A conditional reformulation of $\mathfrak{S}_3$ was called $\mathfrak{S}_4$ and a slight modification of $\mathfrak{S}_3$ in which definitions were of a different form was called $\mathfrak{S}_5$. All those axiomatizations are presented in Sect. 3.5. Section 3.7 is an annotated guide to the secondary literature of the subject.

Chapter 4 is concerned with Leśniewski's Ontology. Section 4.1 introduces the basic intuitions behind the system and indicates in what respects the system differs from the classical logic. In Sect. 4.2 the language of Ontology is described in more detail, and Sect. 4.3 presents the 1920 axiomatization of Ontology. Some examples of definitions of Ontology that highlight the flexibility of the language in question are given in Sect. 4.4. A good example of a reasoning led in Ontology is Leśniewski's argument against universals. A certain formulation of this argument is discussed in Sect. 4.5. I also discuss the relation between Ontology and Russell's description theory in Sect. 4.6 and Leśniewski's attempt to employ Ontology to deal with a certain paradox (having to do with four-dimensional objects) in Sect. 4.7. Secondary literature pertaining to Ontology is discussed in Sect. 4.9.

Leśniewski's Mereology is discussed in Chap. 5. In Sect. 5.1, I focus on Leśniewski's motivations for mereology. Section 5.2 describes and compares the axiomatizations of Mereology from years 1916, 1918, 1920 and 1921. Section 5.3 introduces those theorems of Mereology which initially were intended to support the claim that Mereology can be a sensible replacement for set theory and also those theorems which indicate to what extent Mereology turns out to be different from the standard set theory. In the same section I also critically assess the potential role of Mereology in the foundations of mathematics. Section 5.4 is a survey of secondary literature related to Mereology.

In Chap. 6 Severi K. Hämari and I focus on a persistent myth about Leśniewski and definitions, according to which it was Leśniewski who came up with the consistency and conservativeness requirements on definitions. After some preliminaries (Sects. 6.1 and 6.2) We explain the origins of the folklore (Sects. 6.3 and 6.5), elaborate on Leśniewski's unusual style which contributed to the obscurity of his works (Sect. 6.4), explain why Leśniewski's definitions are creative (section creativity), and what Leśniewski's rules for definitions actually accomplished (Sect. 6.7). Finally, we argue that most of the credit on the Polish ground should go to Łukasiewicz and Ajdukiewicz instead (Sect. 6.8).

Things get more involved with Chap. 7 where I discuss various ways one might attempt to replace standard set theory within the framework of Leśniewski's system. In Sects. 7.1–7.4 I describe and critically assess Leśniewski's and Sobociński's Sobociński (1949b) solution(s) to Russell's paradox. The main point of this part is that the strategy is not successful: there still exists a paradox not noticed by Sobociński— the solution implies that for any two individuals one is an element of the other. In passing, I also discuss the question whether Ontology itself does not contain enough set theory (Sect. 7.5).

Another strategy of dealing with set-theoretic paradoxes, which employs the so-called *higher-order epsilon connectives*, suggested by Lejewski (1985), is discussed in Sects. 7.6 and 7.7. I describe the strategy, apply it to the problems raised by Leśniewski and Sobociński and show that it falls short of providing with a theory strong enough to compete with set theory.

Yet another attempt of turning Mereology into a useful tool in foundations of mathematics, Słupecki (1958) generalized mereology, is presented and assessed in Sect. 7.8. This attempt, however, fails, not only because at certain points it goes against Leśniewski's basic convictions about logic, but also because the resulting system is, in an important sense, not extensional.

In the last Chap. 8 I take a step back. If the Leśniewskian systems are to serve as a tool for the nominalist then even before we ask whether and what theories we can reconstruct within this framework we have to first deal with the question whether the logical systems as they already are do not violate the nominalistic assumptions. Hence, I turn my attention to the question whether Leśniewski's quantifiers binding name variables (or Boolos' plural quantifiers for that matter) commit one to the existence of sets (understood as abstract objects). First, I introduce the language of Quantified Name Logic (QNL) and explain its set–theoretic and substitutional semantics (Sect. 8.2).

Having done that, I discuss the question whether plural quantification commits its users to sets. I start with an argument to the effect that it does (section received). It ultimately rests on the assumption that the logic obtained with the substitutional semantics is devoid of certain theoretically preferable features that QNL with set–theoretic semantics has. I discuss reasons to think that in Sect. 8.4.

Then, I discuss some known attempts to provide Ontology (and thus QNL) with a nominalistically acceptable semantics (Sects. 8.5 and 8.6). I find both these attempts lacking, even though the basic intuitions behind them are quite compelling. In Sect. 8.7 I employ those intuitions to develop a more satisfactory semantics. It is a modal interpretation, where the plural quantifier '$\exists a \, \phi$' (suppose ϕ does not contain free variables other than a) is intuitively read as 'it is possible to introduce a name a which would make ϕ substitutionally true.' There is a proof that QNL with Kripke semantics has the same expressive power as QNL with set–theoretic semantics. In Sect. 8.8 I defend the philosophical viability of this semantics against some known or potential objections.

1.3 Leśniewski's Life and Philosophical Development

Stanisław Leśniewski was born on March 30th, 1886 in Serpukhov, Russia. From 1899 to 1903 he attended gymnasium in Irkutsk. Next, he spent some time traveling and studying. He visited Leipzig (1904–1906, he met Wilhelm Wundt), Heidelberg and Zurich (ca. 1906–1908), and Munich (where he studied at the Ludwig Maximilian University in 1909). In 1910 he moved to Lvov, where within 2 years he completed his Ph.D dissertation under Kazimerz Twardowski's supervision. He spent some time in Warsaw in 1913. Then, he lived in Kimorciszki and Moscow for some time before finally settling in Warsaw, where in 1918 he obtained his habilitation[4] was appointed the Chair of Philosophy of Mathematics at the University of Warsaw. There he supervised his only Ph.D student, Alfred Tarski, who graduated in 1923 Tarski

---

[4] Habilitation is the highest academic qualification in some countries (like Poland and Germany). One usually has to obtain it to supervise Ph.D students and sometimes to teach certain courses.

(1923). He continued teaching at the University of Warsaw till May 13th, 1939, when he died of thyroid cancer.[5]

Leśniewski's first publication was his Ph.D dissertation (1911).[6] the between 1911 and 1916 he published regularly. Most his papers from this period were rather informal, but the problems discussed were related to the philosophy of logic and mathematics. Quite soon he was drawn to the problems raised by Russell's paradox and the debates surrounding the foundations of mathematics. In 1914a he published his first piece on Russell's paradox (where he already employed some mereological intuitions) and his 1916 paper contains the first formulation of his Mereology (a formal theory of parthood). Years 1911–1916 are sometimes referred to as the first period of his philosophical activity.

He did not publish anything between 1916 and 1927. After he moved to Warsaw he taught actively and obtained various results whose publication he withheld. During this period, sometimes dubbed 'the second period of his philosophical development' he worked on developing his systems: Protothetic (at first approximation, a generalization of propositional logic), Ontology (roughly, a higher-order logic) and Mereology.

Eventually, he resumed publication. The main item was his "On the Foundations of Mathematics" series, which consisted of eleven chapters spread over five volumes of the *Polish Philosophical Journal (Przeglą ad Filozoficzny)*. He also published two papers in *Fundamenta Mathematicae* concerned with specifically mathematical problems (axiomatic theories of groups and Abelian groups), and four separate papers where he elaborated on his logical systems, Protothetic and Ontology (Mereology, as a theory of parthood is usually considered an extra-logical theory).

The main events in his biography may be represented by the following chronology:

---

[5] For more details on Leśniewski's biography see for example Betti (2005), Jadczak (1997), Woleński (1985) or Luschei (1962).

[6] All page references to Leśniewski are to the English edition of Leśniewski's *Collected works* published as Leśniewski (1991). The only exception are papers not included in those two volumes.

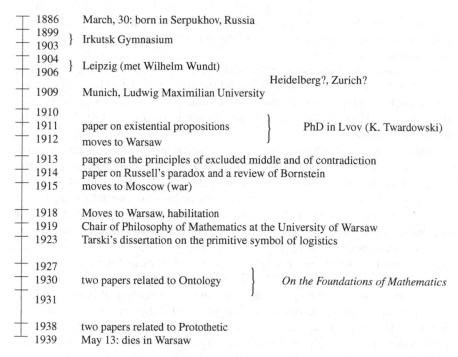

	1886	March, 30: born in Serpukhov, Russia
	1899	} Irkutsk Gymnasium
	1903	
	1904	} Leipzig (met Wilhelm Wundt)
	1906	
		Heidelberg?, Zurich?
	1909	Munich, Ludwig Maximilian University
	1910	
	1911	paper on existential propositions } PhD in Lvov (K. Twardowski)
	1912	moves to Warsaw
	1913	papers on the principles of excluded middle and of contradiction
	1914	paper on Russell's paradox and a review of Bornstein
	1915	moves to Moscow (war)
	1918	Moves to Warsaw, habilitation
	1919	Chair of Philosophy of Mathematics at the University of Warsaw
	1923	Tarski's dissertation on the primitive symbol of logistics
	1927	
	1930	two papers related to Ontology } *On the Foundations of Mathematics*
	1931	
	1938	two papers related to Protothetic
	1939	May 13: dies in Warsaw

As far as we know, from the beginning of his philosophical activity Leśniewski was concerned with philosophy of logic. While in Lvov, he published his dissertation in which he discussed truth and analyticity of existential propositions (Leśniewski 1911). Already in that paper he accepted a somewhat Aristotelian view of the structure of predication. He treated 'is' as a copula which expresses predication and occurs between two terms. This is contrasted with the rendering typical for predicate logic, where one of the terms together with a copula is treated as a predicate, and the other term (the subject term) has to be singular.

Also, from the beginning he decided that mere logic should not distinguish between empty, singular and universal terms (referring to, respectively, no, exactly one, and more than one objects). If one wants to construct a logic to formalize statements of natural language, the intuition behind this decision may be something like this: logic is supposed to provide us with theses and deductions which stay valid when we apply them to natural language. When we use terms in natural language, we quite often do not know (or may be mistaken as to) whether they refer to a unique object or to anything at all (and indeed, answers to such questions are often extra-logical).

For instance, 'Socrates is human', in predicate logic is represented as 'Human(socrates)' where the argument has to be a singular term. The same sentence according to Leśniewski would be rendered as 'Socrates ε human', where (as long as the syntactic correctness of the formula is involved) both 'Socrates' and 'human' can be singular, general or even empty terms. So, for instance, also sentences like 'unicorn ε animal' or 'dog ε animal' would be well–formed, if the language admitted such natural language terms. (This does not mean that they all would be true, but

I will discuss satisfaction conditions of formulas constructed using ε later.) These specific features of his approach will later become trademarks of his logical system of predication called Ontology.

He was also concerned with the validity of the traditional logical principles, which were discussed quite lively in Poland those days. In 1910 Jan Łukasiewicz wrote a book in which he intended to undermine the Principle of Contradiction Łukasiewicz (1910a).[7] Soon after, Leśniewski published a paper (1912) where he criticized Łukasiewicz.

Another principle that raised concerns was the principle of Excluded Middle. Łukasiewicz (1910b) tried to undermine it, and Tadeusz Kotarbiński criticized the principle from the indeterministic perspective. He argued that statements about the future which are not yet causally determined are neither true nor false Kotarbiński (1913). Leśniewski (1913a,b) disagreed, arguing that the only case of failure of this principle is when it is interpreted in a very unusual way and when some terms occurring in a sentence are empty.

Quite soon Leśniewski became interested in paradoxes and the foundations of mathematics. In (1914a) he argued that the Russell class does not exist. While doing so, he tacitly assumed certain principles that hold for mereological fusions, but not for sets in the classical set theory. In (1916), he provided the first construction of his Mereology. This axiomatized theory of the parthood relation, he later suggested, was supposed to constitute an alternative to set theory in the foundations of mathematics.

One of the main themes in Leśniewski's development is his increasing sympathy for nominalism. His early argument against universals (1913b) employed the notion of a property. Later (Leśniewski 1927, 199), he found this argument insufficient and provided a formulation which did not seem to him to refer to properties (whose existence he came to deny). This later formulation is discussed in Sect. 4.5.

Between 1916 and 1927, when he was constructing his systems and did not publish anything, one of his main concerns was to provide a nominalistic description of those calculi. So, instead of providing an inductive definition of an infinite set of well-formed-formulas, he treated formulas as inscription tokens. Instead of speaking of an infinite set of theorems he spoke of a state of a system at a given moment, meaning the set of all inscription tokens which have been actually proven. By an actually proved inscription he meant an inscription token s for which there was a sequence

---

[7] The story is a bit complicated here. First, Łukasiewicz published a book titled *O zasadzie sprzeczności u Arystotelesa. Studium Krytyczne* [*On the principle of contradiction in Aristotle. A critical study*, (Łukasiewicz 1910a)]. In the same year, he published a paper based on this book titled "Über den Satz des Widerspruchs bei Aristoteles" [On the principle of contradiction in Aristotle, (Łukasiewicz 1910c)]. The paper has been translated by Vernon Wedin as Łukasiewicz (1971). Since then, some people when they write about contradiction cite Łukasiewicz (1971) as *the* translation of Łukasiewicz (1910a,b,c) (see e.g. Horn 2009) , as if there was only one Łukasiewicz's work from that year. As of now, the 1910 book hasn't been translated into English. To make things even more confusing, there is a German translation of the 1910 *book* (Łukasiewicz 1993) under exactly the same title as Łukasiewicz's 1910 *paper*. One of the authors who get the story right in their references is Betti (2004b).

of inscription tokens which was a proof of an inscription token equiform[8] to s. This feature made the whole construction cumbersome, especially since another of his points of fixation was precision. He strove to provide a description of the systems in a language as precise as possible—the description he provided is usually referred to as his 'terminological explanations'. (For the sake of accessibility I will not use this method. However, it is interesting for its own sake, and I will touch this issue in Sect. 6.4.)

Leśniewski's nominalistic inclinations also inspired him to put forward Mereology as an alternative to set theory. His main objection against set theories was that they treated sets as abstract objects. He complained that could not even understand what a set as an abstract object is and insisted that the only way that he can understand this notion is when sets are thought of as mereological wholes.

He resumed publishing in 1927. From then on, his published works were mainly concerned with presentation of his logical systems and Mereology. He renounced his early informal philosophical deliberations[9] and focused on formal systems, continuing to work on them till his death.

1.4 Primary Sources

Below is the list of Leśniewski's papers together with English titles and reference to their original place of publication. All page references in this book refer to the two volumes of the English edition of Leśniewski's collected works Leśniewski (1991), even if some translations are slightly corrected. I discuss secondary literature at the end of each of Chaps. 2–5. If the reader is not interested in bibliographical details, they are free to skip ahead to the next chapter. First, there are papers from the period 1911–1916:

1. "Przyczynek do analizy zdań egzystencjalnych" (A contribution to the analysis of existential propositions), (1911).
2. "Próba dowodu ontologicznej zasady sprzeczności" (An attempt at a proof of the ontological principle of contradiction), (1912).

---

[8] The notion of equiformity played an important part. Instead of speaking of an expression type, say ϕ, he explicitly spoke of all tokens equiform to a token ϕ, where ϕ was described in terms of its structure and equiformity of its constituents to certain other tokens. For example, instead of '$p \wedge \neg p$ is not provable in Protothetic', he would say something more like 'for every token whose first and fourth constituents are equiform to the following symbol: p, whose second constituent is equiform to the following symbol: $\wedge$, and whose third constituent is equiform to the following symbol: $\neg$, no system of Protothetic (that is, in no inscribed development state of this system) contains a proof (a sequence of inscription tokens satisfying appropriate conditions) which proves this token.'

[9] In (Leśniewski 1927, 181) he wrote: "Steeped in the influence of John Stuart Mill in which I mainly grew up, and 'conditioned' by the problems of 'universal–grammar' and of logic-semantics in the style of Edward Husserl and by the exponents of the so–called Austrian School, I ineffectually attacked the foundations of 'logistic' from this point of view."

3. "Czy prawda jest tylko wieczna czy też i wieczna i odwieczna?" (Is all truth only true eternally or is it also true without a beginning?), (1913a).
4. "Krytyka logicznej zasady wyłą aczonego środku" (The critique of the logical principle of the excluded middle), (1913b).
5. "Czy klasa klas, niepodporządkowanych sobie, jest podporządkowana sobie?" (Is the class of classes not subordinated to themselves subordinated to itself?), (1914a).
6. "Teoria mnogości na 'podstawach filozoficznych' Benedykta Bornsteina" (A theory of sets based on B. Bornstein's 'philosophical foundations'), (1914b) [not included in Leśniewski (1991)].
7. "Podstawy ogólnej teoryi mnogości I" (Foundations of the general theory of sets I), (1916).

The (1916) paper was the last one written in the early period, and also the first that dealt with the issues discussed in details in the series of papers between 1927 and 1931, titled "On the Foundations of Mathematics I–XI". The list of all parts of this series together with their references is:

1. "O podstawach matematyki. Wstęp. Rozdział I: O pewnych kwestjach, dotyczących sensu tez 'logistycznych'. Rozdział II: O 'antynomji' p. Russella, dotyczącej 'klasy klas, nie będących własnemi elementami'. Rozdział III: O różnych sposobach rozumienia wyrazów 'klasa' i 'zbiór' " (On the foundations of mathematics. Introduction. I. On certain questions concerning the meaning of 'logistic' theorems, II. On Russell's 'antinomy' concerning the 'the class of classes which are not elements of themselves', III. On different ways of understanding the words 'class' and 'set'), (1927).
2. "O podstawach matematyki. Rozdział IV: O podstawach ogólnej teoryj mnogości I" (On the foundations of mathematics. IV. On the foundations of the general theory of sets I), (1928).
3. "O podstawach matematyki. Rozdział V: Dalsze twierdzenia i definicje 'ogólnej teorji mnogości' pochodzące z okresu do r. 1920 włącznie" (On the foundations of mathematics. V. Further theorems and definitions of the general theory of sets up to year 1920), (1929b).
4. "O podstawach matematyki. Rozdział VI: Aksjomatyka 'ogólnej teorji mnogości', pochodzące z r. 1918. Rozdział VII: Aksjomatyka 'ogólnej teorji mnogości', pochodzące z r. 1920. Rozdział VIII: O pewnych ustalonych przez pp. Kuratowskiego i Tarskiego warunkach, wystarczających i koniecznych do tego, by P było klasą p-tów a. Rozdział IX: Dalsze twierdzenia 'ogólnej teorji mnogości', pochodzące z lat 1921–1923" (On the foundations of mathematics. VI. Axiomatization of the general theory of sets from the year 1918. VII. Axiomatization of the general theory of sets from the year 1920. VIII. On certain conditions established by Kuratowski and Tarski, necessary and sufficient for P to be the class of a. IX. Further theorems of the general theory of sets from the years 1921–1923), (1930a).
5. "O podstawach matematyki. Rozdział X: Aksjomatyka 'ogólnej teorji mnogości pochodząca z r. 1921. Rozdział XI: O zdaniach 'jednostkowych' typu '$A\varepsilon b$' "

(On the foundations of mathematics. X. Axiomatization of the general theory of sets from the years 1921–1923. XI. On 'singular' propositions of the type '$A\varepsilon b$'), (1931a).

While working on this series, Leśniewski also published two papers which were more concerned with specific mathematical topics rather than with foundations of mathematics: "Über Funktionen, deren Felder Abelsche Gruppen in bezug auf diese Funktionen sind" (On functions whose fields with respect to these functions are Abelian groups) (1929c) and "Über Funktionen, deren Felder Gruppen mit Rücksicht auf diese Funktionen sind" (On functions whose fields, w.r.t. these functions are groups) (1929d) simplify axiomatizations of two mathematical theories (Abelian group theory and group theory respectively).

In 1929 Leśniewski also published a paper concerned with a description of Protothetic: "Grundzüge eines neuen Systems der Grundlagen der Mathematic § 1–11" (Fundamentals of a new system of the foundations of mathematics, §1–11) (1929a).

In "Über die Grundlagen der Ontologie" (On the Foundations of Ontology) (1930b) he presented an outline of his Ontology and in "Über Definitionen in der sogenannten Theorie der Deduction" (On definitions in the so-called theory of deduction) (1931b) he formalized the requirements he put on definitions in his systems.

In 1938 he was preparing for publication in *Collectanea Logica*[10] two papers: "Einleitende Bemerkungen zur Fortsetzung meiner Mitteilung u.d.T. 'Grundzüge eines neuen Systems der Grundlagen der Mathematik'" (Introductory remarks to the continuation of my article: fundamentals of a new system of the foundations of mathematics) (1938a), where he elaborates on how he had formulated Protothetic and explains some of its basic principles, and "Grundzüge eines neuen Systems der Grundlagen der Mathematik, §12" (Fundamentals of a new system of the foundations of mathematics §12) (1938b) where he hints at another, so-called computative formulation of Protothetic. Those were never published, but preprints survive in the Harvard College Library.

Another invaluable source is a set of notes made by Leśniewski's students. Most of the text consists of formal symbols, theorems and proofs. The notes have been gathered, translated into English and edited by Srzednicki and Stachniak (1988). There are two parts to that book: (a) foundations of mathematics, (b) Peano arithmetic and Whitehead's theory of events. The first part is divided into: "From the Foundations of Protothetic", "Definitions and theses of Leśniewski's Ontology", "Class Theory". The second part consists of: "Primitive terms of arithmetic", "Inductive Definitions", and "Whitehead's Theory of Events".

---

[10] A Polish journal which never came to being, mainly because of the Second World War.

References

Betti, A. (2005). Stanisław Leśniewski - life. The Polish Philosophy Webpage, http://www.fmag. unict.it/ polphil/PolHome.html

Betti, A. (2004b). Łukasiewicz and Leśniewski on contradiction. *Reports on Philosophy*, 22, 247–271.

Cohen, J. (1965). Review: The Logical Systems of Leśniewski by E. *Luschei. The Philosophical Quarterly*, 15(58), 81–82.

Dawson, E. (1965). Review: The Logical Systems of Leśniewski by E. Luschei. *The British Journal for the Philosophy of Science*, 15(60), 341–345.

Gessler, N. (2007). Introduction à l'oeuvre de S. Lesniewski. Fascicule V - Lesniewski, lecteur de Frege. Neuchâtel: Centre de Recherches Sémiologique.

Gessler, N. (2005). Introduction à l'oeuvre de S. Neuchâtel: Lesniewski. Fascicule III - La Méréologie. Neuchâtel: Centre de Recherches Sémiologique.

Grzegorczyk, A. (1955). The systems of Leśniewski in relation to contemporary logical research. *Studia Logica*, 3, 77–95.

Horn, L.R. (2009). Contradiction. In Zalta, E.N. (Ed.), The Stanford Encyclopedia of Philosophy. Spring 2009 edition.

Iwanuś, B. (1973). On Leśniewski's Elementary Ontology. *Studia Logica*, 31, 7–72.

Jadczak, R. (Ed.). (1997). *Mistrz i Jego Uczniowie [Master and His Students]*. Warsaw: Scholar.

Kearns, J. (1973). Review: The Logical Systems of Leśniewski by E. *Luschei. The Journal of Symbolic Logic*, 38(1), 147–148.

Kotarbiński, T. (1913). Zagadnienie istnienia przyszłości [The problem of the existence of the future]. Przegląd Filozoficzny [Philosophical Review], 16, 74–92.

Lejewski, C. (1958). On Leśniewski's Ontology. *Ratio*, 1(2), 150–176.

Lejewski, C. (1985). Accommodating the informal notion of class within the framework of Leśniewski's Ontology. *Dialectica*, 39, 217–241.

Leśniewski, S. (1911). Przyczynek do analizy zdań egzystencjalnych. Przegląd Filozoficzny, 14, 329–245. [A contribution to the analysis of existential propositions, translated as (Leśniewski, 1991, 1–19].

Leśniewski, S. (1912).Próba dowodu ontologicznej zasady sprzeczności Przegląd Filozoficzny, 15, 202–226. [An attempt at a proof of the ontological principle of contradiction, (Leśniewski, 1991, 20–46)].

Leśniewski, S. (1913a). Czy prawda jest tylko wieczna czy też i wieczna i odwieczna? Nowe Tory [New Trails], 8, 493–528. [Is all truth only true eternally or is it also true without a beginning?, (Leśniewski, 1991, 86–114)].

Leśniewski, S. (1913b). Krytyka logicznej zasady wyłączonego środku. Przegląd Filozoficzny [Philosophical Review], 16, 315–352. [The critique of the logical principle of excluded middle, (Leśniewski, 1991, 47–85)].

Leśniewski, S. (1914a). Czy klasa klas, niepodporządkowanych sobie, jest podporządkowana sobie? Przegląd Filozoficzny, 17, 63–75. [Is a class of classes not subordinated to themselves, subordinated to itself?, (Leśniewski, 1991,115–128)].

Leśniewski, S. (1914b). Teoria mnogości na 'podstawach filozoficznych' Benedykta Bornsteina [A theory of sets based on B. Bornstein's 'philosophical foundations']. Przegląd Filozoficzny [Philosophical Review],17, 488–507. [not included in (Leśniewski, 1991)].

Leśniewski, S. (1916). Podstawy ogólnej teoryi mnogości I. Prace Polskiego Koła Naukowego w Moskwie, 2. [Foundations of the general theory of sets I, (Leśniewski, 1991, 129–173)].

Leśniewski, S. (1927). O Podstawach Matematyki, Wstęp. Rozdział I: O pewnych kwestjach, dotyczących sensu tez 'logistycznych'. Rozdział II: O 'antynomji' p. Russella, dotyczącej 'klasy klas, nie będących własnemi elementami'. Rozdział III: O różnych sposobach rozumienia wyrazów 'klasa' i 'zbiór'. Przegląd Filozoficzny, 30:164–206. [On the foundations of mathematics. Introduction. Ch. I. On some questions regarding the sense of the 'logistic' theses. Ch. II. On Russel's 'antinomy' concerning 'the class of classes which are not elements of themselves'. Ch.

III. On various ways of understanding the expression 'class' and 'collection' (Leśniewski, 1991, 174–226)].

Leśniewski, S. (1928). O podstawach matematyki, Rozdział IV: O podstawach ogólnej teoryj mnogości I. Przegląd Filozoficzny, 31, 261–291. [On the foundations of mathematics. Ch. IV On 'Foundations if the general theory of sets. I', (Leśniewski, 1991, 227–263)].

Leśniewski, S. (1929a). Grundzüge eines neuen Systems der Grundlagen der Mathematik §1-11. Fundamenta Mathematicae, 14, 1–81. [Fundamentals of a new system of the foundation of mathematics, §1–11, (Leśniewski, 1991, 410–605)].

Leśniewski, S. (1929b). O podstawach matematyki, Rozdział V: Dalsze twierdzenia i definicje 'ogólnej teorji mnogości' pochodzące z okresu do r. 1920 włącznie. Przegląd Filozoficzny, 32, 60–101. [On the foundations of mathematics. Ch. V. Further theorems and definitions of the 'general theory of sets' from the period up to the year 1920 inclusive, (Leśniewski, 1991, 264–314)].

Leśniewski, S. (1929c). Über Funktionen, deren Felder Abelsche Gruppen in bezug auf diese Funktionen sind. Fundamenta Mathematicae, 14, 242–251. [On functions whose fields with respect to these functions, are Abelian groups, (Leśniewski, 1991, 399–409)].

Leśniewski, S. (1929d). Über Funktionen, deren Felder Gruppen mit Rücksicht auf diese Funktionen sind. Fundamenta Mathematicae, 13, 319–332. [On functions whose fields, with respect to these functions, are groups, (Leśniewski, 1991, 383–398)].

Leśniewski, S. (1930a). O podstawach matematyki, Rozdział VI: Aksjomatyka 'ogólnej teorji mnogości', pochodząca z r. 1918. Rozdział VII: Aksjomatyka 'ogólnej teorji mnogości', pochodząca z r. 1920. Rozdział VIII: O pewnych ustalonych przez pp. Kuratowskiego i Tarskiego warunkach, wystarczających i koniecznych do tego, by p było klasą p-tów a. Rozdział IX: Dalsze twierdzenia 'ogólnej teorji mnogości', pochodzące z lat 1921–1923. Przegląd Filozoficzny, 33, 77–105. [On the foundations of mathematics. Ch. VI. The axiomatization of the 'general theory of sets' fro the year 1918. Ch. VII. The axiomatization of the 'general theory of sets' from the year 1920. Ch. VIII. On certain conditions established by Kuratowski and Tarski which are sufficient and necessary for P to be the class of objects A. Ch. IX. Further theorems of the 'general theory of sets' from the years 1921–1923, (Leśniewski, 1991, 315–349)].

Leśniewski, S. (1930b). Über die Grundlagen der Ontologie. Sprawozdania z posiedzeń Towarzystwa Naukowego Warszawskiego, Wydział Nauk Matematyczno-Fizycznych, 23, 111–132. [On the foundations of Ontology, (Leśniewski, 1991, 606–628)].

Leśniewski, S. (1931a). O podstawach matematyki, Rozdział X: Aksjomatyka 'ogólnej teorji mnogości pochodząca z r. 1921. Rozdział XI: O zdaniach 'jednostkowych' typu 'Aεb'. Przegląd Filozoficzny, 34, 142–170. [On the foundations of mathematics. Ch. X. The axiomatization of the 'general theory of sets' from the year 1921. Ch. XI. On 'singular' propositions of the type 'Aεb', (Leśniewski, 1991, 350–382)].

Leśniewski, S. (1931b). Über Definitionen in der sogenannten Theorie der Deduction. Sprawozdania z posiedzeń Towarzystwa Naukowego Warszawskiego, Wydział Nauk Matematyczno-Fizycznych, 24, 289–309. [On definitions in the so-called theory of deduction, (Leśniewski, 1991, 629–648)].

Leśniewski, S. (1938a). Einleitende Bemerkungen zur Fortsetzung meiner Mitteilung u.d.T. 'Grundzüge eines neuen Systems der Grundlagen der Mathematik'. Widener Library Info Harvard Depository XLL 270.5, Hollis number: 005913328 [Introductory remarks to the continuation of my article: 'Grundzüge eines neuen Systems der Grundlagen der Mathematic', (Leśniewski, 1991, 649–710)].

Leśniewski, S. (1938b). Grundzüge eines neuen Systems der Grundlagen der Mathematik, §12. Widener Library Info Harvard Depository XLL 270.6, Hollis number: 002222243.

Leśniewski, S. (1991). Stanisław Leśniewski. Collected Works (two vols.). Dordrecht: Kluwer Academic Publishers. [Edited and translated by S. Surma, J. Srzednicki, and D. I. Barnett, continuous pagination].

Łukasiewicz, J. (1910a). O Zasadzie Sprzeczności u Arystotelesa - Studium Krytyczne [On the principle of contradiction in Aristotle. A critical study]. Polska Akademia Umiejętności, Warsaw.

[the book is sometimes mistaken with Łukasiewicz 1910b, which is just a paper based on the book].

Łukasiewicz, J. (1910b). O zasadzie wyłączonego środka [On the principle of excluded middle]. Przegląd Filozoficzny, 13, 372–373.

Łukasiewicz, J. (1910c). Über den Satz des Widerspruchs bei Aristoteles. Bulletin international de l'Académie des sciences de Cracovie, 1–2, 15–38.

Łukasiewicz, J. (1971). On the principle of contradiction in Aristotle. Review of Metaphysics, 24, 485–509. [This is a translation of (Łukasiewicz, 1910c) rather than of (Łukasiewicz, 1910a)].

Łukasiewicz, J. (1993). Über den Satz des Widerspruchs bei Aristoteles. Hildesheim: Olms Verlag. [translated by Jacek Barski].

Luschei, E. (1962). The logical systems of Leśniewski. Amsterdam: North-Holland.

Miéville, D. (1984). Un développement des systèmes logiques de Stanislaw Lesniewski. New York: Peter Lang.

Miéville, D. (2004). Introduction à l'oeuvre de S. Lesniewski. Fascicule II: L'ontologie. Neuchâtel: Centre de Recherches Sémiologique.

Miéville, D. (2009). Introduction à l'oeuvre de S. Lesniewski. Fascicule VI: La métalangue d'une syntaxe inscriptionnelle. Neuchâtel: Centre de Recherches Sémiologique.

Miéville, D., & Vernant, D. (1995). Stanislaw Leśniewski aujourd'hui. Neuchâtel: Centre de Recherches Sémiologique.

Miéville, D. (2001). Introduction à l'oeuvre de S. Lesniewski. Fascicule I: La protothétique. Travaux de logique. Université de Neuchâtel.

Peeters, M. (2005). Introduction à l'oeuvre de S. Neuchâtel: Lesniewski. Fascicule IV - L'oeuvre de jeunesse. Neuchâtel: Centre de Recherches Sémiologique.

Quinon, P. (2011). La métalangue d'une syntaxe inscriptionnelle. History and Philosophy of Logic, 32(2), 191–193.

Rickey, F. (1976). A survey of Leśniewski's logic. Studia Logica, 36(4), 407–426.

Słupecki, J. (1953). St. Leśniewski's Protothetic. Studia Logica, 1, 44–112. Included in (Srzednicki and Stachniak, 1998).

Słupecki, J. (1955). St. Leśniewski's calculus of names. Studia Logica, 3, 7–72.

Słupecki, J. (1958). Towards a generalized Mereology of Leśniewski. Studia Logica, 8, 131–154.

Sobociński, B. (1949b). L'analyse de l'antinomie Russellienne par Leśniewski. Methodos, 1–2(1, 2, 3; 6–7):94–107, 220–228, 308–316; 237–257. [translated as "Leśniewski's analysis of Russell's paradox" (Srzednicki and Rickey, 1984, 11–44)].

Srzednicki, J., & Stachniak, Z. (Eds.). (1988). S. Leśniewski's Lecture Notes in Logic. Dordrecht: Kluwer Academic Publishers.

Tarski, A. (1923). O wyrazie pierwotnym logistyki. Przegląd Filozoficzny [Philosophical Movement], 26, 68–69. [Translated as "On the primitive term of logistic" in Logic, semantics, meta-mathematics, Clarendon Press, Oxford 1956, 1–24].

Thomas, I. (1967). Review: The Logical Systems of Leśniewski by E. Luschei. Philosophy of Science, 34(1), 79–80.

Woleński, J. (1985). Filozoficzna Szkoła Lwowsko-Warszawska. PWN: Warszawa. [Translated as Logic and philosophy in the Lvov-Warsaw school, Reidel, Dordrecht, 1989].

Chapter 2
Leśniewski's Early Philosophical Views

Abstract I focus on Leśniewski's papers written before he published formal work on his systems. I begin with a discussion of what Leśniewski called *linguistic conventions*: various postulates introduced to elucidate the meaning and role of certain natural language devices. Then I show how he used them to draw conclusions about existential propositions, the principle of contradiction, the principle of excluded middle, the eternity of truth and the existence of abstract objects. I also explain his approach to paradoxes: Nelson–Grelling's, Meinong's, Epimenides' (the Liar), and Russell's. All Leśniewski's arguments are carefully reconstructed and critically assessed.

2.1 Introductory Remarks

It was not until the late 1920s that Leśniewski started publishing on the systems of logic for which he is famous. Before 1920, he devoted his papers to certain issues in the philosophy of logic and mathematics. However, most of his investigations from that time did not involve any formalized or axiomatized system. He conducted them informally.

Among the problems he discussed were the logical properties of existential propositions, the justification of the Principle of Contradiction, the value of the Principle of Excluded Middle, the eternity of truth, and a few paradoxes: Meinong's, Nelson-Grelling's, Epimenides' and, most importantly, Russell's.

Leśniewski in 1927 apparently repudiated his earlier results (1927, 181–182). Nevertheless, some of his arguments are interesting. Even if not all of them are philosophically compelling, they do cast some light on how his thought developed. Moreover, the reader has the right to assess these early views on his own instead of trusting Leśniewski's self–criticism.

I begin with a discussion of what Leśniewski called his linguistic conventions and I list the basic conventions that he accepted in years 1911–1914. Next, I show how he

R. Urbaniak, *Leśniewski's Systems of Logic and Foundations of Mathematics*,
Trends in Logic 37, DOI: 10.1007/978-3-319-00482-2_2,
© Springer International Publishing Switzerland 2014

applied those conventions to various problems. That is, I discuss what conclusions those conventions helped him to draw about existential propositions, the principle of contradiction, the principle of excluded middle, the eternity of truth, and how he attempted to solve a few paradoxes: Nelson–Grelling's, Meinong's, Epimenides' (the Liar), and Russell's. Finally I say a few words about his rejection of abstract objects.

2.2 Linguistic Conventions

Leśniewski's linguistic conventions are various restrictions he put on natural language to obtain its more regulated version. Those include definitions, which specify how he uses some terms, some general conditions on what the truth conditions of various types of statements are, and some other assumptions about the language of the discourse which are hard to classify.[1]

Leśniewski claimed that if we rely only on common sense, it is hard to see any way out of paradoxes that arise in natural language (like the Liar, which arises when we ask ourselves whether "This sentence is false" is true or false). Commenting on his solution to the Liar (where he employed the assumption than no noun phrase token refers to any expression token of which it is a part) he explained:

> It is also correct to say that the above–mentioned convention is 'arbitrary' in the sense that it conflicts with 'natural intuitions' of language. [...] Since, keeping to 'natural intuitions' of language we get involved in irresolvable paradoxes, these 'intuitions' seem to imply contradiction. The 'artificial' frame of strict conventions is thus a far better instrument of reason than the language dissolving in the opaque contours of 'natural' habits which often imply incurable contradictions—much as the 'artificially' regulated Panama Canal is a better waterway than the 'natural' rapids on the Dnieper. (1913b, 82)

Although he did not use the term 'compositional', one of his main worries about natural language was that in natural language the *symbolic function* (which is Leśniewski's idiosyncratic term for something like 'reference' or 'semantic role', read on for details) of a complex expression is not a straightforward result of the symbolic function of its components. In order to (1) avoid paradoxes that we run into when reasoning in natural language and (2) explain how the symbolic function of a complex expression results from the symbolic functions of its constituents (and the way they are related to each other in this complex expression) Leśniewski introduced some definitions and postulates which were supposed to regulate the language that he used. He called them *linguistic conventions*. Here is how he explained why they are needed:

> The symbolic function of complex linguistic constructions, e.g., propositions, depends on the symbolic functions of the elements of these constructions, that is individual words, and on their mutual relationship. In the unplanned process of development of language,

---

[1] For instance, Tarski-style hierarchization of natural language seems to be a regimentation of the same sort.

the symbolic function of propositions can depend in some particular cases on identical symbolic functions and on identical relationship between specific words—in quite different ways. The planned construction of complex linguistic forms cannot, for representing various contents in the system of theoretical propositions, be confined within the possible results of the unplanned evolution of language. Such construction calls for the formation of certain general conventional-normative schemas to embody the dependence of the symbolic functions of propositions on the symbolic functions of their elements, and on the mutual relationship between these elements. To ascertain whether the given content has been represented adequately or inadequately in a proposition, one has to analyze individually how the speaker's representational intentions relate to the above-mentioned schemas. These schemas should indicate in what way the symbolic function of a proposition should be conditioned [determined R.U.] by the symbolic functions of the particular words and by their mutual relationship. (1911, 16–17)

And also:

I have more than once pointed out that a system of linguistic symbols, just as any other system of symbols, e.g., the system of railway signals, requires the existence of certain rules for constructing the symbols and keys for reading them.[2] I have repeatedly stressed that the functions of various complex linguistic structures, e.g., those of propositions, should depend, in a correctly constructed precise language, upon the functions or the order of particular words—on the basis of certain patterns determined by general normative conventions the knowledge of which permits the correct symbolization of an object in a given language or the decoding of a symbol for a given language. Taking into account the need so specified for a precise language, I established, in my previous papers, various linguistic conventions indicating on what rules the system of linguistic symbols is based and how to understand statements about some constructions which I used in analysis. (1913b, 56)

So, one reason for introducing conventions was to avoid paradoxes. Another one was that the trust we usually have in natural language intuition, though mainly sufficient for everyday communication, does not allow us to avoid ambiguities in more complicated cases and does not provide any method of resolving misunderstandings. Finally, he thought that conventions were needed to ensure the compositionality of symbolic function.

In his early work, Leśniewski did not codify those conventions in a manner typical for an axiomatic system. To a large extent, they are to be found here and there in his papers, sometimes introduced quite ad hoc in order to approach a specific problem. I will list and discuss those conventions in this section, and in the later sections I will show how Leśniewski applied them. Let us begin with some conventions from (1912, 31–42):

(2.1) All expressions are divided into *connoting* and *non-connoting*.

(2.2) An expression is connoting iff it can be defined.[3]

---

[2] Leśniewski's father was a railroad engineer.

[3] "I divide all linguistic expressions into connoting and non-connoting; I adopt the expression 'connoting expression' to denote expressions that can be defined, and the expression 'non-connoting expression' to denote expressions that cannot be defined. The expressions 'man', 'green', 'square circle', 'centaur' are examples of connoting expressions; the expressions 'to a man', 'well', 'at', 'abracadabra', 'object', 'every man is mortal', etc. are examples of non-connoting expressions." (1911, 31)

Leśniewski, at least in the early period, seems to have followed Aristotle in his account of definitions.[4] The way his train of thought develops indicates that 'can be defined' should be read as 'can be defined by means of a classic definition'. A classic definition has the form: '*A* is a *B* which is *C*'. Following Aristotle, *A* here is the defined expression, *B* is called the *genus proximum* (the closest kind) and *C* is a *differentia specifica* (specific difference). *C* has to somehow determine *B*, in the sense that either not every *B* should be *C*, or *C* should be a conjunction of more properties than *B*. Thus, for instance, 'square circle' is a connoting expression, for we can say 'A square circle is a circle which is a square'. In contrast, 'how are you' is not a connoting expression on this view, because for no *B* and *C* we can say: 'How are you is a *B* which is *C*'.

(2.3) All expressions divide into *denoting* and *non-denoting*. Denoting expressions are those which refer to (symbolize) something that exists, while non-denoting expressions are those that do not refer to any existing object. For example, the following expressions denote something: 'man', 'green', 'object', 'the fact that every man possesses the property of mortality', 'every man is mortal'. The relation of denoting is called a symbolic relation. An expression which denotes something is said to possess a symbolic function.[5]

(2.4) If an expression can be used and treated as having a symbolic function, it is said to have a *symbolic disposition*.

The distinction between having a symbolic function and having a symbolic disposition is the difference between actually referring and being able to refer without

---

[4] The similarity between Leśniewski and Aristotle can be observed at least insofar as we speak of the form of a definition:

                                 "Because there is nothing else in the definition besides the primary genus and the differentiae." (Met., Z, 1037^b, 29–30) Aristotle's approach is more metaphysically involved, because he attaches special status to   (the closest kind) and   (specific differences). These notions are strongly connected with his metaphysics.

[5] "I divide all linguistic expressions into those denoting something and those denoting nothing, in other words symbolizing something and symbolizing nothing or expressions which are symbols and those which are not. I call the relation of expressions to the objects denoted (in other words— symbolized) by these expressions, a symbolic relation. I call that property of an expression which consists in its symbolizing something, the symbolic function of that expression. An expression which denotes something, or which possesses the symbolic function, can be exemplified by the following: 'man', 'green', 'object', 'the possessing by every man of the property of mortality', 'every man is mortal', etc. The expressions which do not denote anything, or do not possess symbolic functions, can be exemplified by the following ones: 'abracadabra', 'square circle', 'centaur', 'the possessing by every man of the property of immortality', 'every man is immortal', etc. The expression 'square circle' does not possess a symbolic function because no object is a square circle, in other words there is no such object as could be symbolized by the expression 'square circle'; thus the expression 'square circle' symbolizes no object, in other words symbolizes nothing. The expressions 'possessing by every man of the property of immortality', 'every man is immortal' do not possess a symbolic function because no man is immortal, in other words there is no object that could be symbolized by the aforementioned expressions. Therefore, these expressions symbolize no objects, or symbolize nothing." (1911, 31–32) It is interesting that among denoting expressions Leśniewski included true sentences, but not false sentences. He was not too explicit about what sentences refer to, but it seems that rather than accepting Fregean Truth, he rather took them to refer to facts or relations of inherence between objects and properties.

any change on the part of the language.[6] Empty names and false sentences have a symbolic disposition but not a symbolic function. Names referring to objects and true sentences have both.[7]

(2.5) Connoting expressions having symbolic function divide into expressions for which there is a non-connoting expression which symbolizes the same object, and expressions for which there is no such non-connoting expression. For instance, the expression 'the fact that every man is mortal' denotes the fact that every man is mortal. There is a corresponding non-connoting expression: 'every man is mortal', which denotes the same fact. By contrast, for the connoting expression 'cow' there is no non-connoting expression denoting the same object(s).

(2.6) Accordingly, connoting expressions possessing a symbolic disposition divide into expressions for which there is a non-connoting expressions having the same symbolic disposition, and those for which there is no such connoting expression.[8]

(2.7) Non-connoting expressions for which there is a connoting expression having the same symbolic disposition are called *propositions*. (1911, 34)[9]

(2.8) A proposition is true iff it possesses a symbolic function. Otherwise, it is false. (1911, 34–35)

(2.9) A simple affirmative proposition (that is, a proposition not having a proposition as its proper part) is of the form 'S is P' and it symbolizes the relation of inherence between the object denoted by the subject and the properties connoted by the predicate. An atomic proposition is true if and only if the object denoted by the subject has all the properties connoted by the predicate.[10]

---

[6] I add this reservation, because Leśniewski wouldn't say that, for instance, 'how are you' has a symbolic disposition just because someone can use this phrase to name an elephant.

[7] "I call the property of an expression which consists in that expression's application or treatment (according to, or against, the adopted linguistic conventions) as one possessing the symbolic function, the symbolic disposition of that expression. Thus, e.g., I say that the expressions: 'man', 'hippocentaur', 'every man is mortal', 'the possessing by a hippocentaur of the property of horseness'—possess a symbolic disposition when they are applied, or treated as expression-symbols. The first and third of these expressions possess a symbolic function, but the second and fourth do not because no object is a hippocentaur or the possession by a hippocentaur of the property of horsiness." (1911, 33)

[8] "All connoting expressions possessing a symbolic function can be divided into two groups: expressions which correspond with any non-connoting expression symbolizing the same object, and expressions which correspond with no non-connoting expression symbolizing the same object. Thus, e.g., the connoting expression 'the possessing by every man of the property of mortality' corresponds with a non-connoting expression symbolizing the same object, namely the expression 'every man is mortal'. The latter also symbolizes the possessing by every man of the property of mortality and also the expression 'the possessing by every man of the property of mortality'. Whereas the connoting expression 'man' corresponds with no non-connoting expression which would symbolize the same objects as those symbolized by the word 'man'.

All connoting expressions possessing a symbolic disposition can be divided into two groups: expressions which correspond to any non-connoting expression possessing the same symbolic disposition, and expressions which correspond to no such non-connoting expression."(1911, 34)

[9] Whenever I refer to someone's work, it is my reading of the text, unless quotation marks are present.

[10] "...any proposition possessing a symbolic function symbolizes the possessing by the object, symbolized by the subject of that proposition, of properties connoted by its predicate. This convention

(2.10) "A proposition having a denoting subject and a connoting predicate possesses a symbolic function if[f][11] a singular proposition being its contradictory counterpart does not possess the symbolic function."(1911, 36)

It seems that by 'the relation of inherence between an object and a property' Leśniewski does not mean one and the same relation for all true simple sentences. Rather, he means something like "relation tokens". For him, whenever we have two true atomic sentences such that either their subjects symbolize different objects or the predicates symbolize different properties, the relations of inherence involved are numerically different.

(2.11) "No contradictory proposition possesses a symbolic function."[12] (1911, 36)

(2.12) "If one of two propositions contradicting each other possesses a symbolic function, then the other will not possess one." (1911, 36)

(2.13) A proposition is *true a priori* iff its truth can be demonstrated by means of linguistic conventions alone. It is *false a priori* iff its falsehood can be demonstrated by means of linguistic conventions alone.[13] (1911, 36)

Now, a few conventions to be found elsewhere:

(2.14) "[A] connoting expression W represents any object possessing the properties connoted by the expression 'W'—with the exception of the expression 'W' itself together with those expressions which have at least one element in common with the expression 'W'." (1913b, 64)

Leśniewski was a nominalist (I discuss his argument against universals later on). The above convention about what connoting expressions denote is to be understood as being about what nowadays we would call tokens: utterances and inscriptions. As we will see Sect. (2.8.3), he relied on this move in his response to the Liar paradox.

(Footnote 10 continued)
implies that propositions can symbolize only the relations of inherence."(1911, 36) What is somewhat interesting, Leśniewski, instead of saying that this refers only to simple (atomic) propositions adds a footnote which says: "...I speak of propositions in the sense of the ones reduced to the form of categorical propositions with positive copulas and predicates in the Nominative." (1911, 36) which seems to suggest the claim that all propositions are reducible to propositions of this specific form. This claim however is not essential for further discussion; neither does Leśniewski give a complete set of directions describing how this reduction should proceed for any arbitrary proposition. Therefore, I decided to treat (2.9) as referring to simple propositions only, without assuming the claim about the reducibility of complex propositions (which is implausible anyway).

[11] Leśniewski in his early writings used 'if' in definitions in the sense of 'iff'.

[12] Some conventions I just quote, if they were originally phrased in a concise manner. If I find a simpler and more accessible reformulation, I use it instead.

[13] Probably 'alone' should be read as 'as the only extralogical assumptions.' "I call false a priori all such propositions whose falseness can be demonstrated by means of linguistic conventions alone or the propositions which can be inferred from those conventions. ...On the analogy of the definition of the expression 'proposition false *a priori*', I define the expression 'proposition true *a priori*'. I employ the latter expression to denote such propositions whose validity can be demonstrated by means of linguistic conventions alone or the propositions which can be inferred from these conventions."

(2.15)   A simple affirmative proposition is *analytic* iff it contains no predicates which connote properties that are not connoted by the subject. It is *synthetic* if it contains predicates
which connote (also) such a property (properties) that are not connoted by the subject.

Here is how Leśniewski phrased the distinction:

I use the expression 'analytic proposition' to denote propositions which, being of the form
of propositions with positive copulas, contain no predicates which connote properties that
are not connoted by the subject. I use the expression 'synthetic proposition' to denote those
propositions which, being of the form of propositions with positive copulas, contain predicates which connote also such properties that are not connoted by the subject. Thus, e.g.,
the propositions 'a man has two hands', 'an orphan does not have a mother' are analytic
if we use the word 'man' in the sense of 'mammal with two hands and two legs', and the
word 'orphan' in the sense of 'human being that has neither father nor mother'; it is because
the properties connoted by the predicates of the propositions with positive copulas—'man
is what has two hands' and 'an orphan is what does not have a mother'—that is the properties of having two hands and not having a mother—are connoted by the subjects: 'man'
and 'orphan'. The propositions: 'man creates God in his own likeness' or 'an orphan never
knows a caress in his life', on the other hand, are synthetic because the properties connoted
by the predicates of the propositions with positive copulas: 'man is what creates God to his
own likeness' and 'an orphan is what does not know a caress in its life', that is the properties
of creating God to one's own likeness and of not knowing a caress in one's life, are not
connoted by the subjects. (1911, 2–3)

(2.16)   A proposition with a negative copula ('is not') reduces to a proposition with the
same subject, positive copula and a predicate generated by preceding the predicate of the
previous sentence by 'non', e.g. 'S is not a P' reduces to 'S is a non-P.' [14]

Leśniewski also introduced a convention about proper names which, although not
used afterwards, is interesting in itself:

(2.17)   Proper names connote the property of possessing a name which sounds like a given
proper name. Proper names denote different objects univocally. For example 'James' denotes
every James in the same sense (as far as there is more than one person called 'James').

Since Leśniewski's views on proper names are quite uncommon we will take a closer
look at this view in Sect. 2.3.

2.3 A Digression on Proper Names

Let us start with Leśniewski's formulation of the view expressed in (2.17):

---

[14] "…a proposition with a negative copula can symbolize possessing, by the object denoted by the
subject of that proposition, properties connoted by the expression consisting of the word 'not' and
the predicate of the proposition with a negative copula in question (the negation 'not' is to apply to
the whole expression that follows it). If the proposition with a negative copula has the form: 'no etc
…', then the word 'no' will be substituted, in the process of reduction, by the word 'every'. Thus
the proposition with a negative copula 'no object can both possess and not possess one and the same
property' …symbolizes the possessing by the object denoted by the subject of that sentences …the
possessing by every object of properties connoted by the expression consisting of the word 'not'
and the predicate of the proposition with a negative copula—that is, the expression 'able to both
possess and not possess one and the same property'." (1912, 23)

J.S. Mill says that not all names have connotations. Among those which have no connotations are, according to Mill, proper names such as, e.g., Paul, Caesar on the one hand, and some of the names of attributes on the other. If this were really so, one could foresee certain difficulties in regarding as analytic those positive existential propositions whose subjects are just such names without connotation. Yet even the names which I have mentioned and which according to Mill have no connotation, in my opinion, have connotation; proper names connote the property of possessing a name which sounds like the given proper name, whereas the names of attributes regarded by Mill as lacking connotation, connote either the property of possessing such names, or the property of complete identity with entities which bear such names. Thus, e.g., the name 'Paul' connotes the property of having the name 'Paul', the name 'redness' connotes the property of having the name 'redness'. Instead of 'Paul' we can then say 'a being which has the name 'Paul'', instead of 'redness'—a being which is completely identical with beings that bear the name 'redness'…I shall touch here upon Husserl's thesis that one proper name, e.g., Socrates', can name various objects only because it is ambiguous, just as names such as 'redness'; I do not think this is the case— these names would be equivocal only if, while denoting various objects, they also connoted different properties. In fact the word 'Socrates', while denoting different objects, connotes always one property, that is the property of bearing the name 'Socrates'. [translation slightly corrected R.U.] (1911, 5–6)

This view is not just merely a descriptive theory of proper names. The descriptive theory postulates that for any proper name there is a definite description which uniquely determines its reference. This already assumes that proper names, taken univocally, pick out their referents uniquely. On the other hand, Leśniewski suggested that proper names are not only connotative (as Russell or Frege may be read to have suggested), but also that they are, in a sense, general terms. There are at least three controversial claims about proper names that Leśniewski seems to have been committed to:

(2.18) For any proper name n it is possible that there are at least two different objects, o_1 and o_2 such that if terms n_1 and n_2 refer uniquely to o_1 and o_2 respectively, then both 'n_1 is n' and 'n_2 is n' are true.

(2.19) For any object o, o is a referent of a proper name n iff o has a property of being named by n.

(2.20) The property described in (2.19) is the only property connoted by n.

One might object to (2.18) by saying that whenever we use a proper name, we seem to (or at least we are trying to) refer to a unique object. One can also argue against (2.19) by insisting that it's *prima facie* implausible if we apply possible-world semantics. For indeed, take the name 'Aristotle'. It seems that Aristotle could have been named with a different proper name than the name which he actually was given. Thus, there appears to be a possible world where Aristotle does not possess the property of being named 'Aristotle'.[15] (2.20) seems suspicious because it seems circular: to know what n refers to, I have to know what property it connotes. Yet, to know which object has the property connoted by n I already have to know which object is named by n, and therefore which object n refers to.

---

[15] See (Kripke 1980) regarding related issues.

The objections are not conclusive, though. The fact that whenever we use a noun phrase m we tend to refer to a unique object does not entail that it is not possible that m has more than one referent. This is pretty straightforward and does not require additional arguments. Note also that there are some terms which are general, but which are in the majority of their uses taken as referring to exactly one object (like 'father' or 'mother'). That is, what a term in one of its uses is taken to refer to is not fixed by its connotation only, but also by pragmatic considerations.

Second, proper names in some rare uses are taken to refer to multiple objects. For instance, if a superstitious parapsychologist who believes in "the power of names" says 'Jacob is always impolite' she does not have to refer to any specific person. Rather, she might be understood as saying 'Everyone whose name is 'Jacob' is impolite'. Or, she may say 'Every Jacob is impolite', which seems to be a normal case of quantification and, unless some other reasons against this reading are given, can be treated as such.

It is also sometimes admissible (however uncommon) to use a plural form of a proper name to refer to a few bearers of this name simultaneously. One of languages in which this happens is Polish. For example, if there are two men, each of them being named 'Rafal', you might say (in Polish): 'The Rafals went to the store' if it is clear from the context that those two men are the only Rafals you might be referring to.

The objection against (2.19) is not really lethal either. What it indicates, though, is that (2.19) has to be reformulated. Although the example mentioned in the objection may falsify (2.19) in one of its readings, consider the following intuitive emendation:

(2.21) For any object o, o is a referent of a proper name n (in a language L) iff n has the property of being named by n (in L).

where by 'being named' we speak of some sort of procedure typical for the way proper names are introduced in a given language.

It is true that the object named 'Aristotle' (in English) in the actual world in another possible world is not named 'Aristotle' in any of the languages existing in that possible world. Still, it is true about Aristotle in such possible world that his name is 'Aristotle' in actual English. This indicates that the connotations of proper names on Leśniewski's view are better to be interpreted as not purely descriptive, for the reference to the actual language to which the proper name belongs is rigid.

A more serious concern comes to mind when we ask ourselves how informative (2.17) really is and we doubt whether (2.20) is not circular. Its main claim is that "proper names connote the property of possessing a name which sounds like a given proper name." If this is supposed to explain how names refer, it does not do much more than saying:

[PNA] A proper name denotes the object(s) having the property of being denoted by it.

or

[PNB] The sense of a proper name n (in L) is 'an object denoted by n (in L)'.

On the face of it, both options are problematic. [PNA] is uninformative: it uses the notion of denotation of a proper name to explain what a proper name denotes. [PNB] in an explanation of what sense of a proper name is uses the notion of denotation of a proper name, thus blocking the possibility of using the sense of a name in explaining what it denotes.

Perhaps there are viable ways of facing these challenges. Maybe the notion of denotation on the right-hand side of [PNA] differs from the one on the left-hand side. Maybe an interesting interplay of the notions of sense and (various definitions of) denotation can be developed to show [PNB] says something philosophically plausible. Alas, Leśniewski did not get into any more detail and never returned to the issue.

This sort of approach, dubbed *nominal description theory* of proper names, have been developed in more detail recently by Kent Bach. Bach, most likely independently of Leśniewski, formulated the view on which "when a proper name occurs in a sentence it expresses no substantive property but merely the property of bearing that very name." (Bach 2002, 73).[16] Alas, a longer discussion of Bach's views lies beyond the scope of this book. Let us get back to Leśniewski and his account of existential propositions.

2.4 Existential Propositions

Leśniewski's first publication (1911) was concerned with the properties of existential propositions. This is also the first time he applied his conventions. His claims in that paper were somewhat unusual. On his view, all positive existential propositions are analytic and yet false. Also, some negative existential propositions are analytic and false and some are synthetic, but contradictory and therefore false.

For Leśniewski all existential propositions are atomic. A positive existential proposition is of the form: 'S exists' or equivalently: 'S is a being' (1911, 3). A negative existential proposition has the form: 'S does not exist' or 'S is not a being' where 'being' is a noun phrase meaning the same as Latin *ens*, Greek *on*, or English *object*.

Following convention (2.1) we know that 'being' connotes something only if it can be defined by means of a classic definition. However, to form such a definition, we would have to find such B and C that:

(2.22) [A] being is a B which is C.

would be a classic definition. To make such a definition work, B would have to be a name wider than 'being'. However, (and that is another concept taken from Aristotle)

---

there is no name wider than 'being' or 'object'. Everything that can be named is an object, and as such, a being of some sort.[17]

Thus, on Leśniewski's view, there is no definition of 'being', and the name 'being' does not connote anything. This, together with convention (2.15) entails:

(2.23) All positive existential propositions are analytic.

Recall that according to (2.15) a positive atomic proposition is analytic if and only if its predicate does not connote properties which are not connoted by the subject. Since 'being' does not connote anything, whatever is the subject of a positive existential proposition, its predicate does not connote anything which is not connoted by the subject. At least, this is Leśniewski's argument for this claim.

Nevertheless, every positive existential proposition is, according to Leśniewski, false. His argument is based on convention (2.9). Indeed, '*S* is a being' is true iff *S* has all properties connoted by 'being'. But, since there are no such properties, it is false that *S* has them. (The fact that Leśniewski thought this argument sound indicates that he treated a sentence of the form 'all *A* are *B*' as implying the existence of some *A*'s. If, instead, (2.9) was read as saying that it is true iff no properties connoted by the predicate are not had by the object(s) denoted by the subject, the argument would not go through.)

Now consider negative existential propositions. There are two options. A negative existential proposition is either analytic or it is synthetic.

If '*S*' itself connotes the property of non-being, then '*S* is a non-being' is analytic, and yet false. According to Leśniewski, atomic propositions with contradictory subjects are contradictory themselves, and cannot be true (2.11).

If '*S*' does not connote the property of non-being, '*S* is a non-being' is synthetic and *S* is only said to have the property of non-being. But every '*S*' is reducible to a name phrase 'being which has properties $A_1, A_2, A_3, \ldots$'. So we have a proposition saying something equivalent to:

The being which has properties $A_1, A_2, A_3, \ldots$, is not a being.

which is a contradictory proposition.

Leśniewski's conclusions are unusual. He defended his claims despite their lack of intuitive support. So, first, it might be objected that the definitions that he uses in proving his conclusions are artificial. For Leśniewski, the relevance of a classification

---

[17] "I have said that the predicate of a positive existential proposition which has been brought to the form of a proposition with a positive copula, does not connote anything, except—at most—the property of being greater than one. I maintain this because such a predicate is synonymous with the words 'being' or 'beings' which connote nothing else, even though they denote ('denotation' as used by Mill) everything. This view conflicts with J.S. Mill's theory which says the word 'being' connotes the property of existing. I consider Mill's theory wrong because, should the word 'being' really connote the property of existing, we could define that word as 'that which has the property of existing', or in other words, as a 'being which has the property of existing' (since the definition must indicate not only *differentiae specificae*, but also the *genus*); this would, then, give rise to an inevitable *regressus in infinitum*. The word 'being' cannot be in fact defined at all; the statement that this word does not connote anything is fully in keeping with this fact." (1911, 4–5)

depends on its theoretical usefulness, that is, on whether it allows us either to construct propositions or theories concerning all and only those objects which fall under one of the classes that our classification singles out. And indeed, he thought that his classification makes it easier to put forward some theoretical claims about all and only (simple) analytic propositions and other claims concerning all and only (simple) synthetic propositions. Some examples that Leśniewski gave were: 'All analytic propositions (and only these propositions) contain no such information about the objects symbolized by the subjects, that could not be deduced from the meaning of the subjects' and 'All synthetic propositions (and only these propositions) contain such information about the objects symbolized by the subjects' (1911, 9–10).[18]

One might think, all Leśniewski did is he introduced a few definitions which, together with some assumptions about how the language works, implied some bizarre claims. They sound awkward because the terms employed mean something different from what we usually associate with those terms. Leśniewski, however, treated his conclusions as if they actually held for natural language. Here is what he says about that:

> The idea expressed in the preceding section contradicts current opinions concerning existential propositions: the possibility of constructing both positive and negative, true existential propositions is commonly accepted. Thus e.g., both the proposition 'people exist' and the proposition 'square circles do not exist' are considered true.
>
> I have taken pains to demonstrate the groundlessness of this common attitude towards existential propositions, all that remains for me to do is to cast some light on the probable origin of such a **widespread error**. [emphasis mine] (1911, 15)

He also suggested that his views about analytic propositions were claims about the same group of propositions which was referred to by other logicians and philosophers:

---

[18] About the idea that some classifications are natural and some are artificial he says: "One hears, from time to time, of 'natural' and 'artificial' methods of classification. People who use this form of expression do not usually limit themselves in characterizing particular methods of classification, to the inclusion of these descriptively to either of the above categories; they usually combine such a descriptive characterization of methods of classification with the teleological element of valuation, and they value 'natural' classifications higher than artificial' ones. The origins of the above characterizations of methods of classification, and the positive or negative estimations which accompany these characterizations, can vary immensely from case to case.

Some such cases are determined by various linguistic habits and traditions, others—by more or less well thought out and justified views concerning the problems of theoretical usefulness.

The classification of propositions into analytic and synthetic which I have carried out ...can be, in view of at least one of its consequences, characterized as regarded by some as 'artificial'. Such an 'objection' can in the first place originate from the fact that one of the two classification labels, i.e., that of analytic propositions, comprises two 'very' or 'too' heterogenous groups of propositions: (1) propositions whose predicates connote any properties but not those connoted by the subjects', and (2) propositions whose predicates do not connote the properties connoted by the subjects only because they do not connote any features.

I do not consider it my task to tone down all such dissonances if they arise solely from deeply rooted emotional impulses resulting from some linguistic habits, yet I cannot miss the opportunity to provide my classification of propositions with a 'safety valve' against objections supported by arguments of theoretical usefulness of my classification." (1911, 9)

In modern logic there is a widespread conviction that all analytic propositions are true. Those who advocate such a privileged position of analytic propositions in science claim that the principle of contradiction would be in jeopardy if any proposition could be false in spite of its being analytic [...But] some analytic propositions might be true, while others might be false. [...] Thus, those scholars who hold that all analytic propositions are necessarily true propositions are in error. (1913b, 61–62)

A more moderate conclusion to draw would be that one can take the Kantian definition of analyticity as inclusion of the subject in the predicate quite literally, and that with the aid of Leśniewski's other assumptions, one can argue that this definition does not coincide with the notion of 'being true in virtue of meaning'.[19]

Anyway, Leśniewski goes on to point out that sentences that seem to be true and existential are misrepresented and their "adequate verbal symbols would be non-existential propositions" (1911, 15), that is, instead of '*S* exist(s)', if we take it to be true, we should use 'Some being(s) is(/are) *S*' and by '*S* does not exist' we should use 'No being(s) is(/are) *S*', if we want to represent our beliefs and intentions adequately. His explanation of how he got to this conclusions is somewhat hasty:

The semiotic analysis of the adequacy or inadequacy of certain propositions in relation to the contents which they represent is then ultimately based on a phenomenological analysis of the speaker's representational intentions. (1911, 17)

There is something specific to note about his arguments. His reasoning for the claim that all positive existential propositions are analytic assumes that 'being' does not connote anything (for it is not definable by means of a classic definition). On the other hand, his justification of the claim that negative propositions are false assumes that 'non-being' connotes something. It is far from obvious that by negating a predicate which does not connote something one can get a predicate connoting something.

Arguably, for 'non-being' one cannot construct a classic definition either. For consider:

A non-being is a *B* which is *C*.

We cannot find *B* and *C* to make the above expression true, because (as Leśniewski himself seems to have believed), the truth of the claim that *B* is *C* in a definition requires a *B* which is *C* to exist. A possible objection to this claim is this. It is possible to take any two disjoint *B* and *C* in order to obtain such a definition. The problem with this objection is that Leśniewski used the notion of a classic definition: *B* has to be a *genus proximum* and *C* has to be a *differentia specifica*. That is, *C* has to be included in *B*. Obviously this would be impossible had *B* and *C* been disjoint.

Leśniewski appears to make two of the standard moves that were usually made those days. (1) He suggested that the surface form of a statement does not have to express its content adequately. (2) He used quantificational devices to express existence.

---

[19] For the sake of simplicity, I am putting well-known general concerns about the notion of analyticity aside. See however (Russell 2008) for a defence of analyticity.

Although (1) seems to be a correct observation, Leśniewski does not seem to have been influenced by a similar strategy of Russell's or Frege's. Probably, the extent to which he was acquainted with the western logical thought at that time did not exceed a relatively simple formal theory given by Łukasiewicz in his book on the principle of contradiction (Łukasiewicz 1910).

As to (2), it is unclear which part of Leśniewski's 'Some object is S' does the heavy lifting of existential import. In his early work, Leśniewski was unclear about that. Later on, however, it was the copula 'is' that imported the existence rather than the quantifier.[20]

Leśniewski's general line is this. First, he introduced linguistic conventions and agreed that they do not merely formalize our natural language intuitions. Rather, they are, to some extent, supposed to replace intuitions whose reliability is undermined by paradoxes. Then, he argued that given those conventions it follows that existential propositions do not have certain properties we normally think they do (and that people who share the normal intuitions are wrong). Next, he insisted that to preserve those "normal" intuitions we have to use different sentences to express what we have in mind.

It is not clear why Leśniewski went through all the trouble of playing around with the 'unnatural' reading and insisting that we should change our ways of expressing existential statements in natural language. Perhaps a more viable option would be to say that the logical structure of existential statements is not the one Leśniewski initially attributes to them, but rather the one he introduces to preserve our intuitions about existential statements. The notion of deep logical structure of a natural language sentence, it seems, was not available to Leśniewski at that time, and this might be one reason why he expressed his view as he did. On this re-interpretation, the lengthy discussion of the unnatural reading might be seen as an argument against that interpretation and for the more elaborate quantificational reading of existential statements.

2.5 The Principle of Contradiction

In 1910, Łukasiewicz published a book titled *On the Principle of Contradiction in Aristotle* (Łukasiewicz 1910a)[21] His considerations of Aristotle's position led him to distinguishing at least three different principles: the logical principle of contradiction,

(2.24)   For no p: p and *not* $- p$ are true simultaneously.

the ontological principle of contradiction,

(2.25)   No being may possess and not possess at the same time one and the same quality.

---

[20] So, for example, it is a theorem of one of Leśniewski's systems that $\exists a \, \neg ex(a)$, i.e. 'For some a, a does not exist'.

[21] For a short explanation of the confusion surrounding Łukasiewicz's works on contradiction, see footnote 7 in Chap. 1. For a wider historical context see Betti (2004b).

and the doxastic principle of contradiction.

(2.26) It is not possible to believe in a contradiction.

He argued that (2.26) as a factual statement is false (although it may have some worth as a methodological directive), and that Aristotle's arguments for (2.24) and (2.25) are not conclusive.

In his paper from 1912, Leśniewski, against Łukasiewicz, attempted a proof of the (ontological) principle of contradiction. The basic version in which he formulated that principle was:

(2.27) No being can both possess and not possess one and the same property.

Which was also equivalently stated as:

(2.28) Every object is unable to both possess and not possess one and the same property.

(2.29) Every P is unable to both have and not have c.

The expression 'both to possess and not to possess a property c' Leśniewski abbreviated as 'to be a contradiction'. The expression 'P is unable to be a contradiction' was read by Leśniewski as equivalent to a conjunction of two claims:

(2.30) P is not a contradiction.

(2.31) The sentence 'P is not a contradiction' is true a priori.

where being true a priori, by convention (2.13), means being provable on the basis of linguistic conventions only.

Leśniewski's argument for the Principle of Contradiction goes as follows:

(2.32) 'Some object is a contradiction' is a contradictory proposition.

By (2.11), 'Some object is a contradiction' does not possess a symbolic function (because (2.11) says that no contradictory proposition possesses a symbolic function). However, it has (he claimed) a denoting subject and a connoting predicate. Therefore, by the convention regarding the truth and falsehood of contradicting propositions with denoting subjects and connoting predicates (2.10), the proposition:

(2.33) Every object is not a contradiction.

is true (because its contradictory proposition is false). Since the proof is based on linguistic conventions only, the Principle of Contradiction is, according to Leśniewski, true *a priori*.

Clearly, in some reasonable sense of the word 'proof', what Leśniewski gave was a proof of the ontological Principle of Contradiction. Nevertheless, epistemically speaking the proof is not extremely compelling. The conventions on which it is based, especially the one which says that no contradictory proposition possesses a symbolic function, seem to do the heavy lifting and to smuggle the conclusion in already. If someone rejects the ontological Principle of Contradiction, they probably will not accept the claim that contradictory propositions do not have a symbolic function.

Given that the argument is not too convincing, it is interesting to see how Łukasiewicz reacted to Leśniewski's paper. Here is a passage from Łukasiewicz's entry in his diary from May 9, 1949[22]:

> I met Leśniewski in Lvov in 1912. I lived then with my uncle in Chmielowski Street 10. One afternoon someone rang at the entrance door. I opened the door and I saw a young man with a light, sharp beard, a hat with a wide brim and a big black cockade instead of a tie. The young man bowed and asked kindly: "Does Professor Łukasiewicz live here?". I replied that it was so. "Are you Professor Łukasiewicz?" asked the stranger. I replied that it was so. "I am Leśniewski, and I have come to show you the proofs of an article I have written against you". I invited the man into my room. It turned out that Leśniewski was publishing in Przegląd Filozoficzny [Philosophical Review, RU] an article containing criticism of some views of mine in the "Principle of Contradiction in Aristotle". This criticism was written with such scientific exactness, that I could not find any points which I could take up with him. I remember that when, after hours of discussion, Leśniewski parted from me, I went out as usual to the Kawiarnia Szkocka (Scottish Café), and I declared to my colleagues waiting there that I would have to give up my logical interests. A firm had sprung up whose competition I was not able to face.

2.6 The Principle of Excluded Middle

Leśniewski (1913b) makes a distinction between two formulations of the Principle of Excluded Middle (PEM). One refers to objects (the metaphysical PEM):

> (2.34) Every object has to stand, with respect to every property, either in the relation of possessing it or in the relation of not possessing it.

the second refers to propositions (logical PEM):

> (2.35) At least one of two contradictory propositions has to be true.[23]

He clearly states that his rejection of PEM will be the rejection of the logical version. He speaks of this version of PEM in not very favorable terms:

> The logical principle of the excluded middle not only does not help to resolve 'logical' problems of various kinds, but it is in fact a dangerous theoretical obstacle which should be, therefore, removed from science. (1913b, 47)

The argument against the logical version of PEM proceeds as follows. According to convention (2.9), for an atomic proposition of the kind 'A is B' to be true, there must

---

[22] The diary is unpublished. The translation from Polish is due to Owen LeBlanc and Arianna Betti. Accessed on Agust 31, 2010 athttp://www.segr-did2.fmag.unict.it/polphil/PolPhil/Lesnie/LesnieDoc.html#lukdiary

[23] Although Leśniewski did not give a definition of the contradictory of a sentence, from his discussion it seems that this was a syntactic notion: if S is taken to be a singular term, the sentence contradictory with 'S is P' would be 'S is not P', and if S is taken to be a general term, then the sentence contradictory to 'Some S is P' is 'Every S is not P'.

be an object denoted by A, having all the properties connoted by B (recall that the reading of 'all properties' is strong here, so: 'all properties and at least one property'). Hence, every atomic proposition whose subject does not denote anything is a false proposition, and every proposition whose predicate does not connote anything is a false proposition.

Thus, all contradictory sentences from the following pairs are false:

(2.36)   Every centaur has a tail. Some centaur does not have a tail.

(2.37)   Every square circle is a circle. Some square circle is not a circle.

(2.38)   Pegasus is an animal. Pegasus is not an animal.

The first two examples come from Leśniewski, the last one does not.[24] Since these pairs are, according to Leśniewski, pairs of contradictory propositions, they are counterexamples to PEM.

Thus, Leśniewski rejects the equivalence:

(2.39)   ('A is B' is false) $\equiv$ (A is not B)[25]

Moreover, this account yields some specific examples of false analytic sentences:

(2.40)   Every contradictory object is contradictory.

(2.41)   A square circle is a circle.

Note that according to Leśniewski, the proposition contradictory to 'S is P' is 'S is not a P', which is not equivalent to 'It is not the case that S is P.' To some extent it allows one to think that his discussion is not a discussion of propositional Principle of Excluded Middle proper, because the negation involved is different. Yet, we will later see that the issues discussed in the 1914 paper will be important for his

---

[24] "Let us suppose that I am to answer the question of whether the following propositions are true: 'every centaur has a tail', 'a certain centaur does not have a tail', 'every square circle is a circle', 'a certain square circle is not a circle'. If we take into account the above analysis, the answer to this question becomes quite easy. Each of the four mentioned propositions is obviously false because the subject of each denotes nothing. The word 'centaur' which is the subject in the first two propositions, and the expression 'square circle' being the subject in the remaining two—denote nothing because no object is a centaur and no object is a square circle. Thus, no object is such that it could be denoted only by the word 'centaur' or by the expression 'square circle'. These expressions denote no objects, that is to say—they denote nothing." (1913b, 59)

[25] "The points raised …throw some light on the 'problem' of negative propositions. They demonstrate the falsehood of the theory of negative propositions, developed in considerable detail by Sigwart in his *Logic* and defended by some other modern logicians. According to this theory—the negative proposition 'A is not B' is equivalent to the affirmative proposition 'the proposition 'A is B' is false'…Given the propositions 'the centaur has no tail', 'a square circle is not a circle' …the respective propositions …are 'the proposition 'the centaur has a tail' is false' and 'the proposition 'the square circle is a circle' is false'. The propositions of the type 'A is not B' …are in this case false because …they have subjects which denote nothing …For the same reason, the respective propositions of the type 'A is B', i.e. 'the centaur has a tail', 'a square circle is a circle'—are also false. If, however, the last two propositions are false, then the propositions stating their falsehood must be true." (1913b, 59–60)

logical development further on (for example, for Leśniewski's solution of Russell's paradox).

Also, Leśniewski's considerations allow us to see the difference between the metaphysical PEM and the logical PEM. Leśniewski's rejection of logical PEM does not undermine the metaphysical version. Leśniewski still claimed that any existing object either has or does not have a given property. The only case where he thought the logical PEM might fail is when a subject of a sentence fails to denote.[26]

2.7 Eternity of Truth

Leśniewski believed that any sentence that was true is and will be true, and that any sentence that was false is and will be false. In this sense, he believed in the eternity of truth. Most of his views on this issue can be found in a paper which he wrote to criticize Tadeusz Kotarbiński, who believed that there are some sentences which are neither true nor false and may become true or false with the flow of time. The whole discussion between Kotarbiński and Leśniewski was seminal for Łukasiewicz's invention of three-valued logics. What Łukasiewicz rejected was the claim that if a sentence is not true, it is false. Thus, he introduced a third value, interpreted as 'not yet determined'.[27]

Since Kotarbiński's views are quite important for the development of many-valued logics (his views inspired Łukasiewicz to construct his three-valued logic) and Kotarbiński's paper has never been translated from Polish, I will start with a lengthy presentation of Kotarbiński's views. Then I will turn to Leśniewski's criticism of those views.

In (1913) Tadeusz Kotarbiński, a representative of the Lvov-Warsaw school, published a paper titled "The problem of the existence of the future." His main claim is that propositions about future events not yet causally determined are neither false nor true now. The arguments are rather unclear and the paper employs many metaphors. The main gist, however, is that if propositions do not change their truth-value through time, then all actions are predetermined and there is no free will (and since we know we have free will, propositions change their truth-value through time). Here is a sample of Kotarbiński's style in that paper:

> Truly one has said that what has gone away has never ceased to exist—it only came into absence. What happened, truly happened; one who states that, states the truth, so this some-

---

[26] Even though Leśniewski rejected one of the forms of the PEM, he did not divide sentences into true, false and indeterminate—according to him, if a PEM failure takes place, it is because both a sentence and its negation are false. Gelber (2004) suggested that "on the basis of his interpretation of Aristotle, for instance, Leśniewski developed a three-valued classification with true, false and indeterminate as the values" (p. 231). This is not Leśniewski's view. The charitable reading of Gelber is that Leśniewski is there mistaken with Łukasiewicz).

[27] The introduction of such a third value would invalidate Leśniewski's reasoning. There are some other difficulties to Łukasiewicz account of *futura contingentia*, though. Alas, they lie beyond the scope of this book.

thing does exist. To change the past—this is a mad idea. To give somebody the world of the past under his reign is to yield an enormous sphere of reality which he cannot reign. If impossibilities are prone to gradation, it is more impossible to revert yesterday's flight of a mosquito than to modify the moon's trajectory tomorrow. What has happened cannot be undone. But really, is it only true about the past? Is it only the past things and events which are such and such, so that they cannot be not such and such, or is it also the things that are to come that are already such and such and not different and they cannot come about as not such and such? Is it also the case that also in the future there is something that cannot be undone, because it has already happened? Many would firmly deny that. But after a consideration, having understood the issue, they would equally firmly assert it. And a scientifically educated man, with his eyes shut is eager to concede without a doubt that the whole future is such and such, and even though nothing in it has happened, it nevertheless cannot be undone. What is to come, allegedly differs from what has gone just like, for example, what happens now behind us in space differs from what is happening in front of us … like a cloud above our heads from another cloud above the Pacific Ocean. [My translation of (Kotarbiński 1913: 74)].

Kotarbiński's reasoning resembles the discussion in Aristotle's *Peri Hermeneias*, IX. The main line of the argument is this. Suppose it is true that p is a sentence true at t'. Then it is now true that p is true at t'. But if that is the case, I cannot do anything to change it, because it has already been 'decided' that p at t' is to come. An analogous reasoning seems to apply to false sentences. Suppose that p is a sentence false at t'. Then, it is false that p is true at t'. But if it is already so, there is nothing we can do to change it.

Speaking more in terms of Kotarbiński's terminology, Kotarbiński identifies being an object of a true affirmative judgment with existence ('judgment' is taken from Kotarbiński's terminology). He argues that if a judgement is true at some time t, then it is true at any time later than t. It follows that whatever exists at some time t, exists at any time later than t (i.e. never ceases to exist), and whatever does not exist now, never comes to existence (for if what p states does not exist, then what $not—p$ states does exist and continues do to so).

This conclusion seems to him quite problematic. We usually believe that we can create something: whether it is a result of our moral action, or a subject of our artistic or intellectual activity. Since this (following his definition of existence) cannot be the case (at least he so believes), if the propositions referring to those created objects have already been true before the act of creation, he concludes that some propositions referring to future freely created objects are neither true nor false.

This is just a guess about what the argument would have been like if Kotarbiński formulated it more explicitly. Here is how he originally phrased his view (again, I quote Kotarbiński at length since his work has not been translated):

Truly, an enormous majority of things to come, practically speaking, have more to do with the past than with future things. The flow of sea currents, earthquakes, volcano eruptions, the rotations of heavenly bodies cannot be changed. We cannot make it true that golf stream will flow through Poland, that the tomorrow's golf stream's flow is here. With this sort of things we are powerless. No matter whether we leave them behind or whether they are in the future. We have to reckon with those to come as well as with those that has passed. Those things are such and such and their having those properties in the future exists even though they are in the future …It is already a true judgment which asserts that an object which lives will die in the future. My inherence, pictorially speaking, of the property of being dead is true already. But is my every position in the future equally existent? Is the fact that I will die equally true as the fact that I will die at this specific time, and that this and no other job I will have, that among two roads I will choose the one on the right and not the one on the left, that in that specific moment such and such thought will cross my mind, raised by my attention in that moment, that I will make a vow (or not), abide by it (or not)? Is this also already true today and was it true centuries ago, or at least since my birth? No, never! Those things are *undecided*, they are in our hands, under our reign, and the great practical difference which divides everything in two spheres is not the present moment: it only crosses the present moment at some places. Of course, on the other side of the river that flows on the border of our freedom, there remains the whole gigantic world of things that are ready to come, which already have the truth to accompany them. The freedom ends where the truth begins, not where the past comes into being. If I can do something, create it, it is not true that it already is. For how can one create what is, what happened or what already is created? One may, at most, create something similar, but not the very same thing. But neither what I can create is false. For how can one come to being, if the judgment stating its existence is false? If it came about (as created), we would have had a contradiction: a judgment about it would have been true and false. Apparently, maybe not, but in fact—indeed. For we are talking not about making something which is not present a future thing. We are not talking about a microbe infecting a presently healthy man. We are not talking about covering a canvas which is now white with paint. We do not need creativity (in our ontological sense) for that; and this is what one has in mind when he denies intuitively the impossibility of creating something when the corresponding judgment about it is false. Only when we truly create, we create the truth…For, for something to truly come into being it is required that before it does, the judgment stating it to be not true. This is the condition of creativity. On the other hand, there is no creativity, if before that moment where the judgment is to begin to be true, it is false. For what is false, cannot become true. If it is false that in the moment *t* an object has a given feature, one cannot make it that this object has this feature at *t*, even if *t* is a future moment. Otherwise—a contradiction. For if one can make this happen, then the assertion of this achievement cannot contradict any true judgment, whereas from the assertion of this achievement it follows that the assertion of what is achieved is true, that it is true that the object has this feature in that moment, but the assumption says that it is false. Contradiction. If anyone wanted to argue that the possibility of doing something does not presuppose the existence of the result, but rather only that the result will exist, and that he does not presuppose the truth of the corresponding judgment now, but rather that later, when we perform the action, it will be true, one can answer what follows. If it is false that a given object has a given feature right now, it will never be true, even when the action will have been performed; moreover, it will still be false. So a certain judgment will be true and will not be true (it will even be false), so—contradiction. For contradiction applies to the future, present and past existence equally. So, one cannot leave tomorrow, if we know today that who asserts this departure lies. [My translation of (Kotarbiński 1913: 79–80)]

The most striking feature of Kotarbiński's view is that he identifies truth with necessity and falsehood with impossibility:

Every truth is a necessity, every falsehood—an impossibility. For what is, has to be, because it cannot not be; if something is ready [i.e. already decided], but it is not, it cannot be, because already the judgment stating it is false, so it cannot become a truth. One can memorize it in a simpler way: a judgment is true if the thing is, and therefore when it is necessary, it is false—if the thing is impossible and it contradicts something that is.[28]

---

[28] "Każda prawda jest koniecznością, każdy fałsz—niemożliwością. Gdy co jest bowiem, być musi, bo nie być nie może; gdy coś gotowe jest, a nie jest, to być nie może, bo już sąd o nim twierdzący jest fałszem, a więc nie może się stać prawdą. Można więc prościej sobie zapamiętać: sąd prawdziwy

In response to Kotarbiński, Leśniewski published a paper: "Is all truth only true eternally or is it also true without a beginning?" (1913a). Leśniewski begins with a few minor points.

Leśniewski on Kotarbiński's Account of Existence

Kotarbiński's at some point in his paper says:of existence:

(2.42) x exists if and only if x is an object such that an affirmative judgment referring to it is true.

Clearly, Kotarbiński's approach differs from Leśniewski's. Leśniewski defined truth in terms of the existence of an object named by the subject and the connotation of the predicate. Kotarbiński on the other hand explains what it is to exist in terms of truth.

(2.42) makes the notion of existence only as clear as our understanding of what a judgment refers to. Indeed, this is the notion that Leśniewski picks on. He adds a few premisses of his own:

(2.43) Every affirmative judgment asserts that an object possesses a property (that is, asserts a relation of inherence).[29]

(2.44) Being an object to which a judgment refers is the same as being asserted by this judgment.

and concludes:

(2.45) Only relations of inherence can exist.

This, as he says, contradicts Kotarbiński, who explicitly stated his own existence (and obviously, Kotarbiński is not a relation of inherence).

This does not seem to be a very charitable reading of Kotarbiński, who probably would not accept all these additional premises. True, Kotarbiński sometimes speaks as if judgments referred to facts (or relations of inherence) and sometimes he speaks as if judgments referred to common-sense objects. One option to read (2.42) charitably is to take it to say that (at least for atomic judgments) a subject of a judgment refers if its name is a subject in a true judgment and that the judgment is about this object. Another is to say that a judgment refers both to the inherence relation *and* to the object referred to by the subject. Either way, the objection is not compelling.

On What Has Gone by

Another Leśniewski's argument relies on intuitions he has about what exists. Opposing Kotarbiński's claim that what is gone by has not ceased to exists, Leśniewski (without giving any definition or argument) deploys the intuition that a relation of

---

(Footnote 28 continued)
jest, gdy rzecz jest, a więc gdy jest konieczna, fałszywy—gdy niemożliwa, sprzeczna z czymś, co jest." (Kotarbiński 1913: 88–89)

[29] The word comes from Latin *in*-(English: *in*) and *-haerere* (English: *to hang, to stick*). The relation of inherence is the relation between a property and an object that has it.

inherence exists at t if and only if it actually takes place at t. He then insists that a relation which has gone by (i.e. ceased to take place) no longer exists.

This also is not a lethal objection. Kotarbiński himself mentions at least two different notions of existence:

> Simply, the term in use is ambiguous. In one sense (which we will use) every object about which an affirmative judgment stating it is true, and vice versa: a judgment stating its object is true if the object exists …in the other sense, the more colloquial one, only the present things exist, the past things existed and the future things will exist. (Kotarbiński 1913: 75)

and the objection stems from their conflation.

Objects with No Judgments

Leśniewski's another objection to (2.42) is that an object can exist without a judgment referring to it being true. This is supported by a nominalistic interpretation of Kotarbiński's notion of a judgment: it is possible that an object exists while no sentence-token referring to it exists. The impression one can get is that Kotarbiński and Leśniewski are talking past each other using different notions of a judgement.

Leśniewski's Notion of Eternal Truth

Leśniewski points out that is unlikely that there are eternally existing tokens: inscriptions, utterances etc. However, to be able to speak of eternal truths, he suggests a more charitable sense of 'being an eternal truth' on which 'being true' means 'being true if uttered'. In this sense, Leśniewski claims that all true sentences are true eternally.

One wonders why Leśniewski himself didn't employ this interpretation when criticizing Kotarbiński by saying that an object can exist without any true sentence token about it.

Another point is that (as we shall later see) Leśniewski was an extensionalist (among other things, he believed that the only connectives used in logic should be truth-functional). That being the case, it is hard to see how conditionalizing and saying 'true if uttered' helps, if the 'if' there is interpreted as material implication. For suppose no tokens of a certain type of sentence p and its negation are uttered. Then 'if p is uttered, it is true' and 'if $\neg p$ is uttered, it is true' both come out vacuously true, and so both p and $\neg p$ are eternally true.

Leśniewski: What is True Will Always be True

Let us move to Leśniewski's main argument:

(2.46)   Suppose some judgement 'A is B' which is now true, will be false.

Therefore:

(2.47)   At some time t later than now, 'A is B' is not true.

But:

(2.48)   Since 'A is B' is not true at t, 'A is not B' is true at t.

This, Leśniewski claims, constitutes a contradiction:

(2.49) "This assumption leads then to the conclusion that the judgment '*A* is not *B*', which contradicts the judgment '*A* is *B*' true at the present time, becomes true at time *t*. An obstacle to the acceptance of the above conclusion is presented by the logical principle of contradiction which says that if one of two contradictory judgments is true, then the other must be false. Thus, if the judgment '*A* is *B*' is true at the present time, we must conclude that the contradictory judgment '*A* is not *B*' is always thus also at time *t*, false." (1913a, 97–98)

A possible response (discussed by Leśniewski) is to say that if *t* is some later time, '*A* is *B*' can be true now and false at *t*, because *A* can cease to be *B* some time between now and *t*.[30] To avoid this difficulty, Leśniewski insists that an atomic sentence with a predicate with no rigid time reference can express different claims at different times.

For instance, 'Stanisław Leśniewski will die' uttered in 1913 expresses the statement that at some time later than the time of utterance Leśniewski will cease to exist. The claim is true in 1913. If it is uttered after his death, it states the existence of a different relation of inherence. The former judgment is true, the latter is not (if Leśniewski does not exist, he can bear the relation of inherence to anything). So, Leśniewski focused on sentences whose subjects denote and predicates connote the same independently of the context (especially time) of utterance.

> ...if in any of the above judgments I substitute for the expressions whose meaning varies with time or circumstances such expressions whose semantic function is (for the given system of linguistic symbols) constant, I will immediately encounter truths which are eternally true. If, e.g., instead of judgments: 'Stanisław Leśniewski will die' or 'Stanisław Leśniewski will cease to be alive' I formulate the judgment: 'Stanisław Leśniewski possesses the property of having ceased to be alive in the future of 2 p.m., March 2nd, 1913', then this judgment will be always true." (1913a, 98–99)

Indeed, we may safely assume that both Kotarbiński and Leśniewski did not discuss the eternity of truth of context- (especially tense-) sensitive sentences, ant that the problem they focused on was rather whether those full-blown, precise sentences can change their truth values.

With this in mind, let us try to run Leśniewski's argument without tense-sensitive predicates and with a particular example of a sentence. Suppose 'Leśniewski is alive' is uttered (and true) at time t_1 (say, in 1913). Uttered at t_1 it expresses the judgment: 'Leśniewski is alive at t_1.' Suppose further that this judgment is false at some later time t_2 (say in 1940). This means that at t_2 it is false that 'Leśniewski is alive at t_1'. But then, at t_2 it is true that 'Leśniewski is not alive at t_1'.

Now, why would this be a contradiction? Recall that Leśniewski explicitly says that he relies on "the logical principle of contradiction which says that if one of two

---

[30] "One might say that my proof would be quite valid but for the fault that it does not 'accord with reality'. To support this objection one might cite a 'random', it seems, judgment which is originally true and then ceases to be true, e.g., the judgment 'Stanisław Leśniewski will die'. This judgment is true for as long as I am alive; when I die it will become false because, when I shall not be here anymore, I shall not be able to die again. By becoming false at the time of my death, the judgment 'Stanisław Leśniewski will die' will give way to the true judgment 'Stanisław Leśniewski died' which in its turn is false until I shall die." (1913a, 97–98)

contradictory judgments is true, then the other must be false. Thus, if the judgment '*A* is *B*' is true at the present time, we must conclude that the contradictory judgment '*A* is not *B*' is always [...] false." (1913a, 97–98) Applied to our case, this entails that if 'Leśniewski is alive at t_1' is true at t_1, 'Leśniewski is not alive at t_1' is false at t_1, and therefore always false, also at t_2. But then 'Leśniewski is alive at t_1' is true at t_2, which contradicts the assumption.

Leśniewski: What is True, Always was True

Leśniewski's argument for the claim that all truth is true without a beginning is very similar.

(2.50)   Suppose some judgement '*A* is *B*' which is now true, was false.

Hence:

(2.51)   At some time t earlier than now, '*A* is *B*' was not true.

But:

(2.52)   Since '*A* is *B*' is not true at t, '*A* is not *B*' is true at t.

And again, Leśniewski says:

(2.53)   ...on the basis of the law of contradiction—since the judgment '*A* is *B*' is true at present, its contradiction '*A* is not *B*' is always false, and so it was at time t". (1913a, 103)

Assessment

Both Leśniewski's arguments rely on (at least) two assumptions. The principle of excluded middle (if a sentence is not true at t, then its negation is true at t), and an unusual version of the principle of contradiction which says that if '*A* is *B*' is true at present, its contradiction '*A* is not *B*' is **always** false.

One might object that the principle of excluded middle has been rejected by Leśniewski himself in his earlier paper. This is not a strong objection, because Leśniewski still claimed the principle is valid for sentences without empty or non-connoting names.

A more serious problem was that Leśniewski used a very unusual version of the principle of contradiction. Let us even put aside the fact that it differs from his own formulation from his paper on this principle. What is more worrying is that the eternal aspect of truth, which is not present in standard formulations of this principle, is built into it. What the principle of contradiction validates is the passage from:

'Leśniewski is alive at t_1' is true at t_1.

to

(2.54)   'Leśniewski is not alive at t_1' is false at t_1.

Alas, the claim that (2.54) entails that 'Leśniewski is not alive at t_1' is always false and therefore also at t_2 is neither normally built into the principle of contradiction, nor an assumption that Kotarbiński would endorse.

Leśniewski in his alleged application of the principle of contradiction actually performs two steps. The first one is the actual application of the original principle, and the second one assumes that iterated applications of temporally indexed truth attributions do not matter. That is, that once we have a sentence 'A is B at t_1', adding or removing 'is true at t_2' to it does not change the truth-value of the sentence. But this is exactly what Kotarbiński contested.

2.8 Paradoxes

In the course of discussing the issues described above, Leśniewski approached a few paradoxes. Three of them (Nelson-Grelling's, Meinong's and Epimenides') were discussed in passing in papers concerned mainly with something else, whereas Russell's paradox was discussed specifically in 1914a and 1927.

2.8.1 The Nelson-Grelling Paradox

The paradox[31] consists in asking 'Does a man who kills all and only people who do not commit a suicide kill himself?' and reasoning: If he kills himself, then he is not among the people who do not commit a suicide, so he does not kill himself. If he does not kill himself, then he is among the people who do not commit a suicide, and thus he kills himself.

Leśniewski's answer in (1913b, 74–77) was that both: 'He kills himself.' and 'He does not kill himself.' are false, because there is no object which is a man who kills all and only people who do not commit a suicide. If there was such an object, this object would be contradictory. (Leśniewski's considerations pertaining to the principle of excluded middle are relevant here. It is exactly because the subject is empty why both sentences can be false.)

2.8.2 Meinong's Paradox

Meinong put forward the claim that whenever a judgement is being made, there is something about which this judgement is being made. Here is how he puts it:

> That knowing is impossible without something being known, and more generally, that judgments and ideas or presentations are impossible without being judgments about and presentations of something, is revealed to be self-evident by a quite elementary examination of these experiences. (Meinong 1960, 76)

---

[31] Grelling's paradox is sometimes associated with quite a different reasoning pertaining to heterologicality. However, 'Nelson-Grelling paradox' is a term which Leśniewski originally used. Indeed, this is similar to a formulation to be found in Grelling and Nelson (1908).

This also pertains to negative existential statements, which claim that something does not exist:

> ...it is even more instructive to recall this trivial fact, which does not yet go beyond the realm of the *Seinsobjektiv*: Any particular thing that is not real must at least be capable of serving as the Object for those judgments which grasp its *Nichtsein*. It does not matter whether this *Nichtsein* is necessary or merely factual; nor does it matter in the first case whether the necessity stems from the essence of the Object or whether it stems from aspects which are external to the Object in question. In order to know that there is no round square, I must make a judgment about the round square...Those who like paradoxical modes of expression could very well say: "There are objects of which it is true that there are no such objects". (Meinong 1960, 82–83)

Specifically, Meinong seems to have been committed also to 'Some objects are contradictory'. Leśniewski gave what he thought to be a generalized version of Meinong's argument for this claim. Here it is.

Either it is true that there are no contradictory objects, or it is false that there are no contradictory objects. If the latter is the case, we easily get to the conclusion that some objects are contradictory. Suppose the former:

(2.55)   There are no contradictory objects.

This seems to imply (at least on the Meinongian view):

(2.56)   A contradictory object is not an object.

But 'a contradictory object is not an object' is true only if a certain object is contradictory. Thus we have:

(2.57)   Some object is contradictory.

The point of this paradox is that to deny the existence of contradictory objects one has to truly predicate something about them. True predication requires the existence of the objects of which something is predicated, and thus to be able to say truly about contradictory objects that they do not exist one has to assume that there are contradictory objects.[32]

Leśniewski argues that there is a simple response to the above reasoning. First, in the Leśniewskian framework, the proposition 'A contradictory object is not an object' is false, because its subject does not denote anything (recall that affirmative sentences with empty subjects were for Leśniewski false). Yet, it is true that there are no contradictory objects. Hence, (2.56) (which Leśniewski takes to be false) is not

---

[32] "...if it were true that there are no 'contradictory objects', in other words, no objects are contradictory, then it would be true that 'a contradictory object is not an object'. It can be, however, true that 'a contradictory object' is not an object only in the case when a certain object is 'contradictory'. If no object were 'contradictory', then no proposition about the 'contradictory object' could be true, including the proposition 'a contradictory object is not an object'. Thus, if it were true: 'a contradictory object is not an object', then it must be also true that a certain object is contradictory. This being so, the assumption made at the beginning that no object is 'contradictory' entails the conclusion that a certain object is 'contradictory'." (1913b, 62–63)

just a reformulation of (2.55) (which he takes to be true) and does not follow from it (because no false sentence follows from a true one).

Following Leśniewski's remarks on existential propositions, the non-existence of contradictory objects should be rather parsed as:

(2.58) No object is contradictory.

which is true and does not imply the existence of any contradictory object, in contrast to (2.55) or (2.56).

2.8.3 Epimenides' Paradox (Liar)

The paradox, in Leśniewski's formulation, consists in the following reasoning. Suppose that Epimenides during the time between t_1 and t_2 utters exactly one sentence:

(2.59) The sentence asserted by Epimenides at time t_1-t_2 is false.

Is this sentence, as uttered by Epimenides, true or false? Suppose it is false. Then, what is stated in this sentence does not take place. Hence, it is not the case that the sentence uttered by Epimenides at time t_1 t_2 is false (for this is what the sentence states). But, if it is not false, it is true.

Suppose it is true. Then, what is stated in this sentence takes place. So it is the case that the sentence uttered by Epimenides at time t_1-t_2 is false.

Leśniewski's solution to this paradox makes use of (2.14), which is a somewhat artificial restriction on what names can refer to (namely, that they cannot refer to expressions whose parts they are), and the following claims:

(2.60) Sentences, understood as inscriptions or sounds, are the proper truth bearers.

(2.61) All affirmative categorical sentences with non-denoting subjects are false.

Consider the expression 'The sentence asserted by Epimenides at time t_1-t_2' as uttered by Epimenides at time t_1-t_2 (as it occurred in the sentence which he uttered). Does it denote at all? Following Leśniewski's convention regarding the denotation of connoting expressions, it denotes all and only those sentence tokens asserted by Epimenides at time t_1-t_2, which have no common part with the expression 'the sentence asserted by Epimenides at time t_1-t_2' as uttered by Epimenides in t_1-t_2. But there is no such a sentence. Therefore, this noun phrase does not denote anything.

Knowing that the subject of Epimenides' sentence does not have a denotation, Leśniewski could truly say in 1913:

(2.62) The sentence asserted by Epimenides at time t_1-t_2 is false.

It is an expression equiform to that uttered by Epimenides. Nevertheless, it was true, while the one uttered by Epimenides was false. Why?

Consider the subject of the sentence uttered by Leśniewski: 'The sentence asserted by Epimenides at time t_1-t_2'. The sentence actually asserted by Epimenides has all properties connoted by this subject, and does not have any expression (numerically) in common with this subject. Therefore the subject of the sentence uttered by Leśniewski denotes the sentence uttered by Epimenides. Moreover, since the sentence uttered by Epimenides does not have a denoting subject, it is true (when uttered by someone else than Epimenides, or by Epimenides but not at t_1-t_2) that it is false.

Note also that from the fact that the sentence uttered by Epimenides is false it does not follow that it is true, because its subject does not refer to this sentence itself, and because the principle of excluded middle fails for sentences with non-denoting subjects.

There are some interesting aspects of this solution. First of all, it seems to be one of the very first (in modern logic) applications of some sort of a restriction excluding self-reference to the Liar paradox.

Second, it assumes that truth-value ascription is token-based. Two tokens of the same sentence (type) which do not contain any indexical expressions, uttered at two distinct occasions can have different truth-values.

Third, quite a few approaches to some paradoxes consist in claiming paradoxical sentences to be meaningless or devoid of truth-value. In contrast, Leśniewski's approach makes them false, without making their negations true (and that is where he employs his rejection of the principle of excluded middle).

Fourth, this solution does not save us from some versions of the liar paradox. (2.14) is too weak. Indeed, assume that no name names an expression of which it is a part. This does not undermine for example the following formulation:

(2.63) Sentence (2.64) is true.

(2.64) Sentence (2.63) is false.

No name in (2.63 and 2.64) names an expression of which it is a part. Still, a contradiction is derivable if we assume that every sentence is either true or false. Perhaps, a stronger assumption should have been taken to avoid the liar. Say, something like 'No circularity in naming should take place', that is, if we start with a given name, say m_0, then there are no expressions $m_1, \ldots, m_k$ such that m_i names a part of m_{i+1} and $m_k = m_0$.

One worry about this sort of approach is that it is pretty strong and it seems to restrain us from considering some sentences which seem quite innocently true. This is a general problem with many solutions to the Liar paradox, and it is not clear whether any solution to the Liar can be given that does not violate some of our intuitions.[33,34] Overall, Leśniewski's approach is of the same type as (and predates)

---

[33] See for instance Simmons (1993) for a critical survey.

[34] Another problem is that this does not provide a way out of the Yablo's paradox. Consider an infinite sequence of sentences $s_0, s_1, s_2, \ldots$ such that:

$$s_0 = \text{'}\forall x \, (P_1(x) \rightarrow \neg Tr(x))\text{'},$$

$$s_1 = \text{'}\forall x \, (P_2(x) \rightarrow \neg Tr(x))\text{'},$$

$$s_2 = \text{'}\forall x \, (P_3(x) \rightarrow \neg Tr(x))\text{'}, \ldots$$

(2.65)

Tarskian approach to natural language and paradoxes, and shares its virtues and vices.[35]

2.8.4 Russell's Paradox

Leśniewski's approach to Russell's paradox is quite specific. As he addresses the problem in (1914a) he applies pretty much the same strategy which he applied in his earlier papers. He introduces his own understanding of certain notions and then argues that when they are taken in this sense the solution is such-and-such. When he discusses Russell's paradox, he does preserve its verbal formulation. But as he goes on explaining how his solution is supposed to work, it becomes more and more clear that what he means by 'class' is not what Frege or Russell had in mind.

While Frege's and Russell's notion of a class is closely related to the notion of set as it is employed in standard set theories (like Zermelo-Fraenkel set theory), Leśniewski's notion stems from intuitions about parthood relation. In Leśniewski's sense, a class of certain objects is just one bigger object whose those former objects are all parts of. So, he approaches Russell's paradox from the mereological (*meros* meaning *part* in Greek) perspective.[36] To see this clearly we need to follow Leśniewski's reasoning for a while.

As phrased by Leśniewski, the paradox assumes that one of the following sentences is true:

(2.66) The class of classes not subordinated to themselves is subordinated to itself.

(2.67) The class of classes not subordinated to themselves is not subordinated to itself.

where by 'subordinated to' Leśniewski means 'being an element of'. Next, standard paradoxical reasoning indicates that each of the above sentences implies the other, thus yielding a contradiction. If such a class is not subordinated to itself, it belongs to the class of of classes not subordinated to themselves, that is, to itself. If it is

---

(Footnote 34 continued)

Assume that the extension of every P_n, for $n = 1, 2, 3, \ldots$, is $s_n, s_{n+1}, s_{n+2}, \ldots$. So every s_i says that all s_j's with $j > i$ are not true. Now ask yourself: is s_0 true? If yes, then for any $k > 0$ the sentence s_k is false. But this also means that for any $k > 1$ the sentence s_k is false. But this is exactly what s_1 says and hence s_1 is true, which falsifies s_0. Suppose then that s_0 is false. This means that there is a $k > 0$ such that s_k is true. But we can repeat the reasoning we led about s_0, this time about this s_k to show that s_k can't be true. Hence the paradox. However, no circularity in reference in Leśniewski's sense takes place: the subject of each sentence refers to all sentences below it. The question whether circularity is involved is sensitive to the notion of circularity involved. See for instance Leitgeb (2002) and Urbaniak (2009a).

[35] See Betti (2004a) for a historical discussion of the relation between Leśniewski's and Tarski's approach to the Liar. See for instance Soames (1999) for a more in-depth discussion of the Tarskian approach.

[36] This move will be discussed in detail in Chap. 5. Problems with Leśniewski's solution to Russell's paradox will be also discussed later on (Chap. 7). Here I just focus on presenting the solution from his early writings.

subordinated to itself, it does not posses the property which all its members are supposed to have, and therefore is not subordinated to itself.

Leśniewski, referring to his discussion of the principle of excluded middle, points out that if no object is the class of classes not subordinated to themselves, then both (2.66) and (2.67) are plainly false, without implying a contradiction.

Before we look at the main assumptions of Leśniewski's argument, let us see how he used variables. For the needs at hand it is enough to think about variables substitutionally. The substitution class contains all possible countable noun phrases. The difference between capital and lower case name variables is not essential. When Leśniewski wants to emphasize that the satisfaction of a formula requires that an individual name be substituted for a variable, he uses a capital letter. Since this was not a part of his official notation, I will use this convention in informal considerations, but dispose of it once I move to Leśniewski's formal systems.

So, for instance, 'an object P is subordinated to a class K iff for some a ...' does not contain variables of three different sorts, but rather three different variables of one and the same type. Moreover, Leśniewski did not introduce a separate sort of variables for classes. He just used P and K to keep track of what variable stands for what (P is the first letter of 'przedmiot'—'object' in Polish, and K is the first letter of 'klasa'—'class').

If there is no empty set, it is enough to show that no object is the class of classes not subordinated to themselves in order to conclude that Russell's class does not exist. This is the strategy that Leśniewski employs.

The basic notion used in Leśniewski's early solution to Russell's paradox (1914a) is 'the class of object(s) n' (we will use '$Kl(n)$' in symbols), where n is a noun phrase. His assumptions governing this notion are as follows:

(2.68)   There is no empty class.

Presumably, the intuition here is that a class is a mereological whole composed of those objects which are its parts (we also say, *their mereological fusion*). Since every object is its own part (in this sense),[37] there is no object devoid of parts. That is, there can be no mereological fusion if there is nothing to fuse.

(2.69) $Kl(Kl(n)) = Kl(n)$

Again, the underlying intuitions are mereological. If you fuse a bunch of objects, then the fusion of that fusion is just one and the same object. A pile of a pile of stones (if we admit piling single objects) is just the same pile of stones.

(2.70)   Every atomic sentence with an empty subject is false.

This stems from convention (2.9). For an atomic sentence to be true, the object denoted by the subject has to have the properties connoted by the predicate. If the object does not denote any object, this condition is not satisfied and the sentence is false.

---

[37] This notion of parthood differs from the notion of *proper parthood* which requires the part to be "smaller" than the object of which it is a part.

(2.71) The universe of existing objects is the class of all objects which are not the universe.

This relies on the intuition that there is nothing more to the universe than all the objects that exist in it: fuse all objects and you have the universe.

Another notion Leśniewski uses is 'subordination' (nowadays, we would say 'being an element'). He considers three ways of defining it and insists that these are the only three options:

(2.72) An object P is subordinated to a class K iff for some a ('for some substitution of a possible countable noun phrase') the following two conditions are fulfilled: (1) K is a class of objects a, (2) P is (an) a (or: P is one of the as).

(2.73) An object P is subordinated to a class K iff for every a the following two conditions are fulfilled: (1) K is a class of objects a,[38] (b) P is a.

(2.74) An object P is subordinated to a class K iff for every a: K is a class of objects a iff P is a.

Leśniewski starts with rejecting options (2.73) and (2.74) from the above trilemma.

Rejecting (2.73)

Suppose that (2.73) works. Take a to stand for 'a square circle'. For no P it is true that P is a square circle. Therefore, there is a name ('a square circle') which, substituted for a always falsifies both conditions formulated in (2.73).

The first condition is falsified because on Leśniewski's view there is no empty class and the class of square circles would have to be empty (and simple sentences with empty subjects are false).

The second condition is false because nothing is a square circle. Thus, for any P and K it is not the case that for every name a: K is the class of objects a and P is a. Thus, on this definition, no object is subordinated to any class.

(Come to think of it, (2.73) is quite absurd to start with. For if K is to be a class of any objects, (2.73) requires it to be the class of a's for any a. Among other things, this would mean that K would have to be both the class of elephants and the class of things which are not elephants.)

Rejecting (2.74)

Suppose (2.74) is true and an object P is subordinated to a class K. Take:

(2.75) a means 'P or not P'.

Clearly:

---

[38] I translate using the indefinite article: 'a class of objects', because the assumption is not taken to imply the uniqueness of a class of objects a.

(2.76) For any object Q, Q is P or not P.

Thus:

(2.77) P is a.

is obviously fulfilled.

Now, recall we use the definition on which:

K is a class of objects a iff P is a.

Since every object is a (on our fixed reading of a), K (the class of objects a) is the universal class:

(2.78) $K = Kl(a)$ is the universal class.

By (2.69) we get:

(2.79) $K = Kl(Kl(a))$

(2.74) also entails:

(2.80) K is a class of the universal class iff P is the universal class.

(just substitute 'the universal class' for a).

(2.79) tells us that K is a class of the universal class, so we get the right-hand side of (2.80):

(2.81) P is the universal class.

On the other hand, replace a from the formulation of (2.74) by 'an object which is not the universal class':

(2.82) K is a class of the objects which are not the universal class iff P is an object which is not the universal class.

Then (since, by (2.71), the universal class is the class of objects which are not the universe) again:

(2.83) Kl(the objects which are not the universal class) is the universal class

Since from (2.78) we already know that K is the universal class, we can conclude that the first condition from the formulation of (2.74) is satisfied. Hence, we also have the other one:

(2.84) P is an object which is not the universal clas.

which contradicts (2.81).

Since the above argument (which follows Leśniewski closely) is somewhat cumbersome, let us take a look at another, less Leśniewskian formulation. (2.74) tells us that for any a:

K is a class of objects a ($K = Kl(a)$) iff P is a.

This allows us to take (2.85), (2.86) and (2.87) as assumptions. Let us proceed with the proof. ('b or non-b' is a complex noun phrase naming all objects which are either b or not b (that is, all objects), 'V' stands for 'the universal class', and 'non-V' is a name which names all things which are not the universal class.)

(2.85) $K = Kl(b$ or non-$b)$ iff P is (b or non-b)   (assumption)

(2.86) $K = Kl(V)$ iff $P = V$ (assumption)

(2.87) $K = Kl(\text{non-}V)$ iff $P \neq V$ (assumption)

(2.88) P is (b or non-b)   (logic)

(2.89) $Kl(b$ or non-$b) = V$ (logic)

(2.90) $K = Kl(b$ or non-$b)$ (2.85), (2.88)

(2.91) $K = V$ (2.89), (2.90)

(2.92) $Kl(Kl(b$ or non-$b)) = Kl(b$ or non-$b)$ (2.69)

(2.93) $Kl(Kl(b$ or non-$b)) = K$ (2.90), (2.92)

(2.94) $Kl(Kl(b$ or non-$b)) = Kl(V)$ (2.89)

(2.95) $K = Kl(V)$ (2.93), (2.94)

(2.96) $P = V$ (2.86), (2.95)

(2.97) $V = Kl(\text{non-}V)$ (2.71)

(2.98) $K = Kl(\text{non-}V)$ (2.91), (2.97)

(2.99) $P \neq V$ (2.87), (2.98)

Contradiction: (2.96), (2.99).

Dealing with (2.72)

Given Leśniewski's assumptions, the only remaining possibility is (2.72). This formulation has two interesting consequences:

(2.100) Every object n is subordinated to $Kl(n)$.

(2.101) Not every object subordinated to $Kl(n)$ is an n.

The first claim is quite simple. For an object n to belong to $Kl(n)$ it is enough that for some a, n is a and $Kl(n)$ is $Kl(a)$. This, however, is witnessed by n itself, for n is n and $Kl(n)$ is $Kl(n)$. (The latter claim is not logically trivial, for on Leśniewski's view, for $Kl(n)$ to exist, n has to exist.)

The second claim, which at first seems unusual, can serve as an example of how 'being subordinated to' is (tacitly, at this point) connected with 'being a part'.

Consider a sphere Q. As Leśniewski says, the class of (all) halves of the sphere Q is the sphere Q itself. The sphere Q is also the class of (all) quadrants of Q. Thus, we have the conclusion that any half of Q is subordinated to the class of quadrants of Q. It is a general point to which we will get back when discussing Leśniewski's understanding of class (Sect. 5.1). Given two names such that the whole constituted by all objects denoted by the first is also the whole constituted by all objects denoted by the second, any object denoted by the first name is subordinated to the class of objects denoted by the second name, even if no object denoted by the first name is denoted by the second name.

Now we are ready to give the argument for the main thesis, saying that every class is subordinated to itself. Define '$\cup$' as a name-forming operator defined by: P is $(Q \cup R)$ iff $(P$ is $Q) \vee (P$ is $R)$. We will show that if K is $Kl(n)$ for some n, then K is subordinated to itself.[39]

(2.102) $K = Kl(n)$ (for some n)   (assumption)

(2.103) $Kl(n) = Kl(n) \cup Kl(n)$ (logic)

(2.104) $Kl(n) = Kl(Kl(n)) \cup Kl(n)$ (2.69), (2.103)

(2.105) $Kl(a) \cup Kl(b) = Kl(a \cup b)$ (for any a, b)   (logic)

(2.106) $Kl(Kl(n)) \cup Kl(n) = Kl(Kl(n) \cup n)$ (2.105)

(2.107) $K = Kl(Kl(n) \cup n)$ (2.102), (2.104), (2.106)

(2.108) K is $Kl(n) \cup n$ (2.102), logic

(2.109) K is subordinated to K iff for some a, $K = Kl(a)$ and K is a (2.72)

(2.110) $K = Kl(Kl(n) \cup n)$ and K is $(Kl(n) \cup n)$ (2.107), (2.108)

(2.111)   For some a, $K = Kl(a)$ and K is a (2.110)

(2.112) K is subordinated to itself   (2.109), (2.111)

This completes Leśniewski's early solution to Russell's paradox. If every class is subordinated to itself, then no class is not subordinated to itself. So the class of classes not subordinated to itself would be empty. But (Leśniewski insists) there is no empty class. Therefore, the class of classes not subordinated to itself does not exist. That being the case, both 'the class of classes not subordinated to itself is subordinated to itself' and 'the class of classes not subordinated to itself is not subordinated to itself' are false. (Since the subject is empty, the principle of excluded middle fails.)

One objection which Leśniewski put forward against the standard formulation of Russell's paradox was that the inference from:

   K is subordinated to the class of classes which are not subordinated to themselves.

to:

   K is not subordinated to itself.

---

[39] This is a streamlined version of Leśniewski's argument, the original is less accessible.

is invalid. In fact, if we assume the mereological account of classes, we think of the class of objects P as a whole consisting of all of P's parts. Moreover, as we will later explain in more detail (in Chap. 5), being an element of (or being subordinated to) this class does not force an object to be a P itself. It may as well be a part of a P or a whole consisting of a few of P's or a whole consisting of some parts of some P's.

For example, consider a herd of ten cows: $a_1, a_2, a_3, \ldots, a_{10}$. Let us use the name 'cow α' as a name which can be truly predicated only of our cows. Now, '$Kl(cow(\alpha))$' is a name of the mereological fusion of our cows. Being an element of a herd is, following Leśniewski, the same as being a part of it. But in this sense, not only our cows are subordinated to this class. Among parts of the fusion are also cow halves, legs, heads, pairs of cows etc. Clearly, being subordinated to the class $Kl(cow(\alpha))$ does not imply being a cow.

Note that assumption (2.71), used in Leśniewski's rejection of (2.74), also shows some predilections towards the mereological understanding of the class. For take the name 'all objects that are not the universal class'. The class of all objects that are not the universal class on the distributive (non-mereological) reading is not the universal class. For there is some object which (by (2.74), which seems to mirror distributive intuitions adequately) is not its element (namely, the universal class).

On the other hand, on the mereological reading the class of a is the mereological fusion of all objects a. This being the case, in some sense, the universal class is nothing over and above the fusion of all other objects in the world. The universal class is a part of the fusion of all objects that are not the universal class.

Leśniewski returned to this paradox after developing his logical systems. A classic discussion of this formal treatment is to be found in Sobociński (1949b). Problems with this approach are discussed in Urbaniak (2008a). I will also discuss this approach later, in Chap. 7.

2.9 On Universals

Leśniewski devoted some space to the rejection of the existence of abstract objects (1913b, 50–53). This reasoning starts with a tentative definition of a general object:

(2.113) A is general with respect to a group of objects iff it possesses only those properties which are common to all those objects.

For instance, the universal 'man' cannot have any properties common only to a few men, but rather it has to have only those properties that all of them have. The argument now proceeds as follows.

Suppose there is an object P_k which is general with respect to individuals $P_1', P_2', \ldots, P_n'$. For every individual object P_i' one can always find certain property c_i which is not common to all individual objects $P_1', P_2', \ldots, P_n'$. For instance, the property of being identical with P_i'. So, P_k does not possess the property c_i.

Moreover, the individual object P_i' does not possess the property of not possessing the property c_i. Hence, the general object P_k does not possess the property of not possessing the property c_i (because, on the currently entertained definition, a general object cannot possess any property which an object that falls under it does not have). But then, P_k possesses the property c_i. Thus, any general (universal) object is contradictory.

The argument is not extremely convincing. The anti-nominalist can simply rely on a different account of universals. For instance, one may make a distinction between the properties of the universal itself (like: being universal, being immaterial, being abstract etc.) and the content of the universal. The universal 'man' can be said to be abstract whereas no individual man is an abstract object. Introducing the notion of content to the account of universals seems to refute Leśniewski's objection. Sure, if an object does not have the property of not possessing the property c_i, then it possesses the property c_i itself. But it is doubtful that if the property of not possessing the property c_i is not in the content of a universal, then the property of c_i is in its content. With respect to a given content properties are not divided in two groups: those, which belong to the content, and those, whose negated properties belong to the content. Rather, they divide in three groups: those which belong to the content, those the negated properties of which belong to the content, and those which are not 'decided' by the content (that is properties such that neither themselves nor their negated properties belong to the content).

Gryganiec (2000) suggests another reason to treat the definition used by Leśniewski as inadequate. Here is how he formulates his argument:

> The incorrectness of the definition of the object Op can be also shown in a different way. Suppose there are two individual objects P_1 and P_2. Assume moreover that the first one, that is P_1, possesses n properties, whereas the second one—P_2—$n + m$ properties. We do not decide here whether symbols n and m denote finite or infinite sets of properties; we only know that those sets exhaust the ontic furnishing of those objects. The universal $Op^{\{P1,P2\}}$ with respect to objects P_1 and P_2— according to Leśniewski's definition—will only have the properties common to objects P_1 and P_2. Therefore, the object $Op^{\{P1,P2\}}$ will have n properties. This being the case, $Op^{\{P1,P2\}}$—thanks to the principle of extensionality—will be identical with the object P_1, which also possesses n properties, that is, it will be an individual. [my translation]

Let's grant that by n and m Gryganiec understands sets of properties and not the cardinalities of those sets (he is somewhat ambiguous about it, but if we do not grant it, the argument obviously will not work).

Still, the argument is not very compelling. First, Gryganiec assumes that for P_1 and P_2 their universal exists. But neither Leśniewski's argument nor the Platonist position requires that for any two objects there be an objects which is their universal. But this is a minor scratch. More importantly, Gryganiec asks us to assume that there exist two distinct individual objects one of which has all the properties that the other has (and some more).But such a possibility is hard to imagine.

If one admits that being identical with an object is a property, and that not having a property is also a property, then take the property of being identical with P_2 (call it f). P_1 clearly does not have property f. Then P_1 has the property of not having the property f. But since P_2 has all the properties of P_1, P_2 also has the property of not having the property f. That is, P_2 has the property of not having the property that it obviously has, which does not seem possible. If, on the other hand, not having a property is not treated as a property, Leśniewski's argument does not work anyway and there is no point in putting forward a counterexample.

We can try to salvage Gryganiec's argument by denying that being identical with an object is a property. If so, we need a stronger and yet reasonable notion of property. Now, the problem is that spatio-temporal location seems to be a good candidate for a property of an individual even in this stronger sense. But then, P_2 could not have all the properties that P_1 has unless they completely overlapped. But if overlapping non-identicals is the metaphysical construct that has to be conjured to make the argument work, the argument is at least controversial.

Leśniewski later formulated a formalized version of his argument. The new version did not employ the notion of a property and was expressed in the language of Ontology. I will get back to this argument in Sect. 4.5 after introducing the required formal apparatus.

2.10 Further Readings

As for Leśniewski's philosophical views, the standard place to start is Luschei (1962). For a reader who wishes to see the character in a wider perspective, (Woleński 1985) (in Polish) is the classic—a good English translation of it is Woleński (1989). He has a chapter on Leśniewski, which as far as technical aspects of the logical systems are concerned is much more detailed than Luschei's book, but is not meant to be comprehensive. Other background-covering works are Woleński (1986a and 1999). The Polish school of logic is presented in a more general survey on the development of logic between the two world wars in Grattan-Guinness (1981). Jadczak (1993a,b) discusses Leśniewski's role in the Lvov-Warsaw school. Leśniewski's letters to Kazimierz Twardowski have been recently published as Leśniewski (1999). The question whether Leśniewski was a philosopher at all is asked (and answered positively) by Woleński (2000). Simons (2008a) provides a very readable survey in his Stanford Encyclopedia entry on Leśniewski.

Sanders (1996) discusses informally Leśniewski's motivations for his systems. He does not go beyond the content of Luschei (1962) and Poli and Libardi (1999). A good but quite informal survey of Leśniewski's achievements is Kearns (1967), which also contains an interesting evaluation of Leśniewski's work. Kearns argues that Leśniewski's view of the world is "defective in that it does not recognize structure" (p. 88), which, he suggests, deprives "Mereology of significance". Leśniewski's early philosophical views are discussed in French by Peeters (2005).

Rickey (1976, reprinted as Srzednicki and Stachniak 1998) provides a more technical overview of some results about Leśniewski's systems obtained before 1972. An interesting survey of technical aspects of Leśniewski's work, especially of the relation between Ontology and set theory has been published by Surma (1977).

Leśniewski's view on truth-bearers in the context of the Lvov-Warsaw school is presented in Woleński and Rojszczak (2005).

Sinisi (1966) discusses Leśniewski's criticism of Whitehead's theory of events.

Betti (2004a) presents Leśniewski's solution to the liar paradox and argues that it was Leśniewski and not Tarski who suggested regimentation of natural language as a device which is necessary to avoid paradoxes.

Regarding paradoxes, an interesting account which employs some of Leśniewski's ideas is Hiż (1984). The relation between the early and informal solution to Russell's paradox from 1914 and the later development of Leśniewski's formal systems are explored by Sinisi (1976).

Leśniewski's nominalism is nicely discussed by Hintze (1995), and the connection between it and Goodman's approach is explored by Prakel (1983). Simons provides a good introduction to Leśniewski's nominalistic inscriptional approach to metalogic (2002) and provides a good background on nominalism in the Lvov-Warsaw school (1993, reprinted in Srzednicki and Stachniak 1998).

Tadeusz Kotarbiński, Leśniewski's good friend, used Leśniewski's systems as a tool while putting forward his own reism (Kotarbiński 1929). As Woleński (1986b) convincingly explains, Ontology itself is neutral with respect to the problem and it is rather the form of the logical system which makes it easier to speak of reducing various non-concrete terms to other terms. In other words, Leśniewski's systems do not contain terms intended to represent abstract entities, but if there were such entities, it would be possible to express facts about them in Leśniewski's languages.

There is also an interesting connection between Brentano's analysis of categorical propositions and Ontology, explored in Simons (1984).

An important discussion of the impact of Leśniewski's thought can be found in Betti (2008b). While Tarski shared nominalistic intuitions with Leśniewski and Tarskian impossibility of giving a satisfactory theory of truth for ordinary language and the analysis of quotation marks can be already found in Leśniewski, there are deep discrepancies between the interests of Tarski and Leśniewski. While the latter was not interested in model theory, the former was partially responsible for its origination. While the former enjoyed working on set theory and metalogic, the latter focused on purely axiomatic work on logical systems, without doing much metatheory and definitely not liking set theory. Yet, Betti convincingly argues that the story about a great breakup between the two logicians is a myth: Tarski's interest were quite different to start with and it seems to her "that the whole story was more, from Tarski's point of view, a fight for freedom from a 100 percent genius master, one whose commitment to a radical philosophical position was, for an extraordinarily gifted and ambitious mathematician, very much in the way." (Betti 2008a, 56)

A somewhat surprising similarity between Leśniewski and Mally has been discovered and described by Gombocz (1979).

Leśniewski's systems are also known for their application to history of philosophy. Thom (1986) interestingly attempts to cast new light on Parmenides, Gorgias, Leucippus and Democritus using a modified version of Ontology. Henry (1972, 1969) attempted to apply Leśniewski's logics to analysis of some medieval concepts.

References

Bach, K. (2002). Giorgione was so-called because of his name. *Philosophical Perspectives*, *16*, 73–103.

Betti, A. (2008a). Polish axiomatics and its truth – on Tarski's Leśniewskian background and the Ajdukiewicz connection. In Patterson(2008) (pp. 44–71).

Betti, A. (2008b). Polish axiomatics and its truth: On Tarski's Leśniewskian background and the Ajdukiewicz connection. In D. Patterson (Ed.), *New Essays on Tarski and Philosophy* (pp. 44–71). Oxford: Oxford University Press.

Betti, A. (2004b). Łukasiewicz and Leśniewski on contradiction. *Reports on Philosophy*, *22*, 247–271.

Betti, A. (2004a). Lesniewski's early liar, Tarski and natural language. *Annals of Pure and Applied Logic*, *127*(1–3), 267–287.

Gelber, H. G. (2004). *It Could Have Been Otherwise. Contingency and Necessity in Dominical Theology at Oxford* (pp. 1300–1350). Leiden-Boston: Brill

Gombocz, W. (1979). Lesniewski und Mally. *Notre Dame Journal of Formal Logic*, *20*, 934–946.

Grattan-Guinness,. (1981). On the development of logic between the two world wars. *The American Mathematical Monthly*, *88*(7), 495–509.

Grelling, K.,& Nelson, L. (1907/1908). Bemerkungen zu den Paradoxien von Russell und Burali-Forti. *Abhandlungen der Fries'schen Schule (Neue Serie)*, *2*, 300–334

Gryganiec, M. (2000). Leśniewski przeciw powszechnikom [Leśniewski against universals]. *Filozofia Nauki*, *3–4*, 109–125.

Henry, D. (1972). *Medieval Logic and Metaphysics: A Modern Introduction.* London: Hutchinson.

Henry, D. (1969). Leśniewski's Ontology and some medieval logicians. *Notre Dame Journal of Formal Logic*, *10*(3), 324–326.

Hintze, H. (1995). Merits of Lesniewski type nominalism. *Logic and Logical Philosophy*, *3*, 101–114.

Hiż, H. (1984). Frege, Leśniewski, and information semantics. on the resolution of antinomies. *Synthese*, *60*, 51–72.

Jadczak, R. (1993b). Stanisław Leśniewski a szkoła Lwowsko-Warszawska. *Analecta: studia i materiały z dziejów nauki*, *2*(2),29–37.

Jadczak, R. (1993a). Pozycja Stanisława Leśniewskiego w szkole Lwowsko-Warszawskiej. *Ruch Filozoficzny*, *50*(3), 311–316.

Katz, J. (1990). Has the description theory of names been refuted? In *Meaning and Method: Essays in honour of Hilary Putnam* (pp. 31–62). Cambridge : Cambridge University Press.

Kearns, J. (1967). The contribution of Leśniewski. *Notre Dame Journal of Formal Logic*, *8*, 61–93.

Kotarbiński, T. (1913). Zagadnienie istnienia przyszłości [The problem of the existence of the future]. *Przegląd Filozoficzny [Philosophical Review]*, *16*,74–92.

Kotarbiński, T. (1929). *Elementy Teorii Poznania, Logiki Formalnej i Metodologii Nauk [Elements of the Theory of Knowledge, Formal Logic and Methodology of the Sciences]*. Lwów: Ossolineum.

Kripke, S. (1980). *Naming and Necessity*. MA: Harvard.

Leitgeb, H. (2002). What is a self-referential sentence? critical remarks on the alleged (non-)circularity of Yablo's paradox. *Logique & Analyse*, *177–178*, 3–14.

Leśniewski, S. (1911). Przyczynek do analizy zdań egzystencjalnych. *Przegląd Filozoficzny*, *14*, 329–245. [A contribution to the analysis of existential propositions, translated as (Leśniewski, 1991, 1–19)].

Leśniewski, S. (1912). Próba dowodu ontologicznej zasady sprzeczności. *Przegląd Filozoficzny*, *15*, 202–226. [An attempt at a proof of the ontological principle of contradiction, (Leśniewski, 1991, 20–46)].

Leśniewski, S. (1913a). Czy prawda jest tylko wieczna czy też i wieczna i odwieczna? *Nowe Tory [New Trails]*, *8*,493–528. [Is all truth only true eternally or is it also true without a beginning?,(Leśniewski, 1991, 86–114)].

Leśniewski, S. (1913b). Krytyka logicznej zasady wyłączonego środku. *Przegląd Filozoficzny [Philosophical Review]*, *16*, 315–352. [The critique of the logical principle of excluded middle, (Leśniewski, 1991, 47–85)].

Leśniewski, S. (1914a). Czy klasa klas, niepodporządkowanych sobie, jest podporządkowana sobie? *Przegląd Filozoficzny*, *17*,63–75. [Is a class of classes not subordinated to themselves, subordinated to itself?, (Leśniewski, 1991, 115–128)].

Leśniewski, S. (1927). O Podstawach Matematyki, Wstęp. Rozdział I: O pewnych kwestjach, dotyczących sensu tez 'logistycznych'. Rozdział II: O 'antynomji' p. Russella, dotyczącej 'klasy klas, nie będących własnemi elementami'. Rozdział III: O różnych sposobach rozumienia wyrazów 'klasa' i 'zbiór'. *Przegląd Filozoficzny*, *30*,164–206. [On the foundations of mathematics. Introduction. Ch. I. On some questions regarding the sense of the 'logistic' theses. Ch. II. On Russel's 'antinomy' concerning 'the class of classes which are not elements of themselves'. Ch. III. On various ways of understanding the expression 'class' and 'collection' (Leśniewski, 1991, 174–226)].

Leśniewski, S. (1999). *Listy do Kazimierza Twardowskiego. Filozofia Nauki*, *6*(1–2), 115–133.

Łukasiewicz, J. (1910a). *O Zasadzie Sprzeczności u Arystotelesa - Studium Krytyczne [On the principle of contradiction in Aristotle. A critical study]*. Polska Akademia Umiejętności : Warsaw. [the book is sometimes mistaken with (Łukasiewicz, 1910c), which is just a paper based on the book].

Luschei, E. (1962). *The Logical Systems of Leśniewski*. Amsterdam: North-Holland.

Meinong, A. (1960). On the theory of objects (translation of 'über Gegenstandstheorie', 1904). In R. Chisholm (Ed.), *Realism and the Background of Phenomenology* (pp. 76–117). Glencoe: Free Press

Peeters, M. (2005). *Introduction à l'oeuvre de S*. Centre de Recherches Sémiologique: Lesniewski. Fascicule IV - L'oeuvre de jeunesse. Neuchâtel.

Poli, R., & Libardi, M. (1999). Logic, theory of science, and metaphysics according to Stanislaw Lesniewski. *Grazer Philosophische Studien*, *57*, 183–219.

Prakel, J. (1983). A Leśniewskian re-examination of Goodman's nominalistic rejection of classes. *Topoi*, *2*, 87–98.

Rickey, F. (1976). A survey of Leśniewski's logic. *Studia Logica*, *36*(4), 407–426.

Russell, G. (2008). *Truth in Virtue of Meaning: a Defence of the Analytic/Synthetic Distinction*. Oxford: Oxford University Press.

Sanders, J. (1996). Stanisław Leśniewski's logical systems. *Axiomathes*, *3*, 407–415.

Simmons, K. (1993). *Universality and the liar. An Essay on Truth and the Diagonal Argument*. Cambridge : Cambridge University Press.

Simons, P. (2008a). *Stanisław Leśniewski*. In Stanford Encyclopedia of Philosophy.

Simons, P. (1984). A Brentanian basis for Lesniewskian logic. *Logique et Analyse*, *27*, 297–308.

Simons, P. (1993). Nominalism in Poland. In F. Coniglione, R. Poli, & J. Woleński (Eds.), *Polish Scientific Philosophy: The Lvov-Warsaw School* (pp. 207–231). Amsterdam: Rodopi.

Simons, P. (2002). Reasoning on a tight budget: Lesniewski's nominalistic metalogic. *Erkenntnis*, *56*(1), 99–122.

Sinisi, V. (1966). Leśniewski's analysis of whitehead's theory of events. *Notre Dame Journal of Formal Logic*, *7*(4)

Sinisi, V. (1976). Lesniewski's analysis of Russell's antinomy. *Notre Dame Journal of Formal Logic*, *17*, 19–34.

Soames, S. (1999). *Understanding Truth*. New York : Oxford University Press.

Sobociński, B. (1949b). L'analyse de l'antinomie Russellienne par Leśniewski. Methodos, 1–2(1, 2, 3; 6–7),94–107, 220–228, 308–316; 237–257. [translated as "Leśniewski's analysis of Russell's paradox" (Sobociński and Rickey, 1984, 11–44)]

J. Srzednicki & Z. Stachniak (Eds.). (1998). *Leśniewski's Systems*. Dordrecht: Protothetic. Kluwer Academic Publishers.

Surma, S. (1977). On the work and influence of Stanisław Leśniewski. In R.O. Gandy and J.M.E. Hyland (Eds.), *Logic Lolloquium 76, Studies in Logic and the Foundations of Mathematics* (pp. 191–220). North-Holland.

Thom, P. (1986). A Lesniewskian reading of ancient ontology: Parmenides to Democritus. *History and Philosophy of Logic, 7,* 155–166.

Urbaniak, R. (2008a). Leśniewski and Russell's paradox: Some problems. *History and Philosophy of Logic, 29*(2), 115–146.

Urbaniak, R. (2009a). Leitgeb, "about", Yablo. *Logique & Analyse, 207,* 239–254.

Woleński, J. (1985). *Filozoficzna Szkoła Lwowsko-Warszawska.* PWN: Warszawa. [Translated as Logic and philosophy in the Lvov-Warsaw school, Reidel, Dordrecht, 1989].

Woleński, J. (1986a). *Filozofia Szkoły Lwowsko-Warszawskiej [Philosophy of the Lvov-Warsaw school].* Wydawnictwo Uniwersytetu Wrocławskiego.

Woleński, J. (1989). *Logic and Philosophy in the Lvov-Warsaw School.* Dordrecht : Kluwer Academic Publisher.

Woleński, J. (1999). *Essays in the History of Logic and Logical Philosophy.* Jagiellonian University Press.

Woleński, J. (2000). Czy Leśniewski był filozofem? [Was Leśniewski a philosopher?]. *Filozofia Nauki, 31–32*(3–4),

Woleński, J., & Rojszczak, A. (2005). *From the Act of Judging to the Sentence: The Problem of Truth Bearers from Bolzano to Tarski.* New York: Springer.

Woleński, J. (1986b). Reism and Leśniewski's Ontology. *History and Philosophy of Logic, 7,* 167–176.

Chapter 3
Leśniewski's Protothetic

Abstract Prototethic is Leśniewski's generalized system of propositional calculus: one is allowed to use quantifiers binding propositional variables and variables representing various connectives or higher-order function variables. First, I describe the historical context of its development, discuss the crucial notion of semantic categories and explain Leśniewski's original notation. The notation uses wheels and spokes to graphically represent connectives, so that geometric relations between connectives correspond to certain logical relations. Then I move on to Leśniewski's motivations for Prototethic (especially his dissatisfaction with *Principia Mathematica*). Since Prototethic resulted from gradual generalizations of the propositional calculus, I describe all those generalizations and provide some examples of proofs. Finally, I include an annotated guide to the secondary literature of the subject.

3.1 Introduction

Leśniewski's first step in his logical investigations was a construction of a set theory (1916) (his theory of parthood called *mereology*) meant "to contribute as much as possible to the justification of modern mathematics" (1916, 129). He did not fully formalize it—he used variables, introduced axioms and definitions, and proved theorems, but except for the use of variables, all these were given in a regimented version of the natural language.[1] All his publications on Mereology (1928,1930a,b,1931a) shared this level of formalization. Only around 1930, quite in parallel, he started publishing on the logic underlying his Mereology:

---

[1] Later, he commented: "In the year 1915 I made a first attempt to give a deductive form to my conception of classes and collections, which I was using in the analysis of the 'antinomy' of Russell. Relying on four propositions called axioms, and using a number of other propositions which I called definitions, I arrived at a number of interesting theorems with methods which I was then able to use." (1928, 227)

R. Urbaniak, *Leśniewski's Systems of Logic and Foundations of Mathematics*,
Trends in Logic 37, DOI: 10.1007/978-3-319-00482-2_3,
© Springer International Publishing Switzerland 2014

While using colloquial language in scientific work and attempting to control its 'logic', I endeavored to somehow rationalize the way in which I was using in colloquial language various types of propositions passed down to us by 'traditional logic'. While relying on 'linguistic instinct' and the often non-uniform tradition of 'traditional logic,' I attempted to devise a consistent method of working with propositions which were 'singular', 'particular', 'general', 'existential' etc. ...Having acquired considerable experience in the consistent scientific operation with these 'singular' propositions[2] and having at my disposal a considerable number of what were for me thoroughly reliable theoretical syntheses which, although they were constructed more or less *ad hoc* and not formulated within the limits of a deductive system, nevertheless made it much easier for me to elucidate for myself and others the subtleties of my scientific language, I wished to take one step further and, using the 'singular' propositions of the type 'A ε b,' to base all my deliberations on some clearly formulated axiomatization which would harmonize with my theoretical practice in this domain at that time. (1931b, 366–367)

The system resulting from this investigations, which dealt mainly with predication was called 'Ontology'. Leśniewski formulated the first single axiom of this system in 1920, but for a while he did not publish his results.[3]

Simultaneously, he started working on a theory that coped with the propositional aspect of his reasonings. In 1921, inspired by Łukasiewicz's modification of Nicod's system of propositional logic (Lukasiewicz 1921) he postulated two things: (1) that his propositional calculus should have only one primitive logical constant, and (2) that there should be no additional symbol for definitions, but rather that definitions should be written using that primitive term.[4]

In 1922 Tarski, Leśniewski's only Ph.D. student, proved that if the universal quantifier binding propositional variables is admitted, the equivalence sign suffices as the only sentential connective in a functionally complete propositional calculus.[5] Since then, they cooperated on the construction of both a theory of propositional deduction (called *Protothetic* from 'first theses' in Greek, loosely speaking) and of a theory which may be roughly characterized as dealing with predication (called *Ontology*).[6]

This chapter will be devoted to the development of the propositional construction. Protothetic originated as the result of gradual axiomatic generalization of the propositional calculus. The first propositional system (Ϭ) given by Leśniewski is a

---

[2] i.e. propositions of the form 'a is b'.

[3] "This theory has for some time now become known to a wider circle of my colleagues and students through copies of my university lectures."(1930b, 608)

[4] "In defining the functions of the theory of deduction in terms of other such functions, both Sheffer and Nicod use a special equal-sign for definitions which they do not define in terms of the primitive functions of the system .... This circumstance makes it difficult to say whether Nicod's theory of deduction is in fact constructed out of the single primitive term: '|'.

"In 1921 I realized that a system of the theory of deduction containing definitions 44 would actually be constructed from a single term only if the definitions were written 45 down with just that primitive term and without recourse to a special equal-sign for 46 definitions." (1938b, 418)

[5] A propositional calculus is called functionally complete (in the relevant sense) if all extensional propositional connectives can be defined by means of the primitive symbols of its language.

[6] Most of the work was done by Leśniewski, but quite often he employs results obtained by Tarski, crediting him explicitly.

purely equivalential propositional calculus without quantifiers. The second system ($\mathfrak{S}_1$) admitted quantifiers binding both propositional variables and variables representing propositional connectives. Extending $\mathfrak{S}_1$ with a certain rule (rule η) results in a stronger system $\mathfrak{S}_2$. However, rule η was quite complicated, so it was replaced by another, simpler rule (called the rule of extensionality, $\eta\star$). The result of this replacement is called $\mathfrak{S}_3$. A conditional reformulation of $\mathfrak{S}_3$ was called $\mathfrak{S}_4$ and an equivalential modification of $\mathfrak{S}_3$ (definitions were written differently) was called $\mathfrak{S}_5$.

There are at least two difficulties in giving a historically adequate account of these systems. The first one is that sources are sparse. Many important manuscripts have been destroyed during the Second World War. The systems are reconstructed mainly on the following basis: Tarski's dissertation (Tarski 1923), the first eleven paragraphs of Leśniewski's "Fundamentals of a New System of the Foundations of Mathematics" (Leśniewski 1929),[7] and some lecture notes taken by his students, edited by Srzednicki and Stachniak (1988).[8]

The second difficulty is that the standards of exposition that Leśniewski followed were quite different from what we are used to nowadays. That is, there is no explicit definition of a language, rules are sometimes described quite vaguely and sometimes only mentioned without explicit statement. The closes that his presentation gets to a precise formulation is in his semi-formalized Terminological Explications, whose accessibility is problematic for other reasons (see Sect. 6.4).

I will begin with a brief account of the semantic categories insofar as this notion is useful for describing Leśniewski's systems and the formal languages he uses. Next, I will devote some time to Leśniewski's notation of propositional connectives. Then, I will present the six propositional calculi. Having done this, I will provide some examples of proofs in Protothetic. Finally, I will survey secondary literature pertaining to Protothetic.

3.2 Semantic Categories and the Language of Protothetic

The idea of segregating expressions of a given language according to the role they play can be traced back at least to Aristotle (first five chapters of *On interpretation*). A theory providing such a division has been developed in late medieval scholastic logic. In the modern era the key philosopher to discuss this issue seems to have been Husserl. He outlined a sketch of the theory of *Bedeutungskategorien* (categories of meaning) in his *Logical Investigations*.

---

[7] The twelth paragraph being prepared for publication in 1938 survived as (Leśniewski 1938b).

[8] Before the notes were published, they were accessible to very few of Leśniewski's followers. One of them, J. Słupecki provided an impressive reconstruction Protothetic in his (1953). An unpublished dissertation titled *Computative Protothetic* by Owen Le Blanc (1991) is an invaluable source of information about computative formulations of Protothetic (Leśniewski (1938b) somewhat hints at the computative formulation).

The first formalized theory of semantic categories was constructed by Leśniewski around 1921. The first mention of this fact seems to be the second footnote of Tarski (1923), but Leśniewski employed his categories in 1929 Leśniewski (1929) as an alternative to Russell's and Whitehead's theory of types. This study has been developed by (Ajdukiewicz 1935), another logician from the Lvov-Warsaw school. However, from the publication of "On syntactic connectivity" (Ajdukiewicz 1935) onward, the subject of this theory was rather called 'syntactic categories'.

The underlying idea is that two expressions belong to the same semantic category if they are interchangeable in any sentence without the loss of meaningfulness, i.e. if one is replaced by another in a meaningful sentence, the resulting expression is also a meaningful sentence.

In all of Leśniewski's systems there are two basic semantic categories: names and sentences. For Leśniewski, a name is any expression (complex or not) that may take the place of either M or N in a sentence of the form: 'M is N'.[9] This is quite straightforward in Latin or Polish which do not have articles.[10] For example, all of the following are well-formed in Latin[11]: 'Canis est canis', 'Canis est animal', 'Socrates est animal', 'Chimera est animal', 'Chimera est Chimera' and therefore 'canis' (dog), 'animal', 'Socrates' and 'Chimera' are all names on this account. Beside expressions of those two categories there are syncategorematic expressions whose role is to construct more complex expressions when applied to some expressions. They are called *functors*.[12] We will need a piece of notation from Ajdukiewicz (1935), called *index notation*.

Definition 3.1 (*Semantic Category, index notation*) Every propositional variable or propositional constant belongs to one and the same class called the semantic category of sentences. If an expression belongs to this category, its index is s. Every name variable or name constant belongs to one and the same category, called the semantic category of names. If an expression belongs to this category, its index is n. Every expression of a language whose index is neither s nor n is called a *functor*. What semantic category it belongs to is determined by (1) how many expressions it is applied to in order to construct a well-formed expression, (2) what the semantic categories of the expressions it is applied to are, and (3) what the semantic category of the whole expression which it generates properly concatenated with its arguments is. If a functor f applies to l arguments: $\alpha_1, \ldots, \alpha_l$ of categories (respectively) of the

---

[9] The fact that this account can be found in a fairly standard textbook used those days in Poland (kotarbinski 1929) indicates that Leśniewski's notion is by no means unusual.

[10] Ultimately, I will use the notion of semantic category with reference to formalized languages only, where related issues do not arise.

[11] It is sometimes believed that in Latin 'est' has to occur as the last word in sentences like these. However plausible it is with respect to certain forms of ancient Latin, the rule does not apply to Medieval Latin, which was the language that logicians used to work in for centuries. The scholastic grammar allowed infix notation for 'est' and I see no reason not to use scholastic Latin in contexts like the present one.

[12] The term comes from Kotarbiński.

following indices: $\sigma_1, \ldots, \sigma_l$ and yields an expression β of the semantic category denoted by σ, the index of the semantic category of f is the fraction: $\frac{\sigma}{\sigma_1 \cdots \sigma_l}$. $\square$

Here is an example of how this works for the language of classical propositional logic:

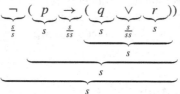

Expressions of (various systems of) Protothetic can be built from: propositional variables, sentence-forming functors of k ($k \in \mathbb{N}$) sentence arguments (where either equivalence or implication is the only primitive connective, depending on the system), the universal quantifier, any constant of any semantic category if it is correctly defined in a given system by means of symbols which have been previously introduced (what a correct definition is will be explained later). Whenever a category has been introduced, variables can be introduced which belong to this category and quantifier can bind them. One way to put it more formally is this:

Definition 3.2 Expressions of Protothetic are constructed from:

- Propositional variables: p, q, r, possibly with numerical subscripts. They are of semantic category s.
- Functor variables: for any category τ constructed from s's there are variables $f^\tau, g^\tau, f_1^\tau, f_2^\tau, \ldots, g_1^\tau, g_2^\tau, \ldots$ of this category.[13] Functor variables can be subscripted with variables or constants of lower semantic categories (in which case they are called *functor variables with parameters*).
- One primitive propositional connective of category $\frac{s}{ss}$: $\equiv$ (in some formulations: $\rightarrow$).
- The universal quantifier: $\forall$ (binding variables of all categories).[14]
- For any category τ there are constants of this category $c_1^\tau, c_2^\tau, \ldots$, defined according to the rules of definition (to be introduced). Constants can be subscripted with variables or constants of lower semantic categories (in which case they are called *constants with parameters*). I will discuss the role of parameters in Sect. 3.6.

---

[13] Strictly speaking, Leśniewski didn't use category superscripts. Nor will we, assuming that the context will make it clear what category a variable is supposed to belong to.

[14] Leśniewski used brackets to mark quantification. In general, the way Leśniewski introduced his language would be somewhat foreign to the modern reader. The presentation I use does not follow Leśniewski to the letter. However, I believe that in the present context the gain in readability outweighs the philosophical advantages of employing Leśniewski's notation and his inscriptional syntax. See however remarks in Sect. 6.4.

The formation rules are:

- If ϕ is an expression of Protothetic of category $\frac{\tau}{\sigma_1\cdots\sigma_n}$ and expressions of Protothetic $\alpha_1,\ldots,\alpha_n$ are respectively of categories $\sigma_1,\ldots,\sigma_n$, $\phi(\alpha_1,\ldots,\alpha_n)$ is a well–formed expression of Protothetic of category τ.
- If χ^τ is a variable of category τ and ϕ is an expression of Prothetic of category s, then $\forall\chi^\tau\ \phi$ is an expression of Protothetic of category s.   □

3.3 Leśniewski's Notation

Leśniewski introduced his own idiosyncratic notation for one- and two-place sentential connectives. Its particular feature was that the shape of a symbol encoded its truth-table. In this section I will describe this notation due to its own merits, but later on I will use a modern notation for the sake of accessibility.[15]

First of all, this is a prefix notation. Functors are written before arguments, arguments are placed in parentheses and are not separated from each other by commas. Universal quantification binding variables of any semantic category is expressed by the whole sequence of those variables put within the lower corner-brackets: ⌊ ⌋. For instance, a universal quantifier binding all propositional variables $p_1,\ldots,p_n$ is: ⌊$p_1,\ldots,p_n$⌋. The scope of a quantifier is marked by upper-corner brackets: ⌈ ⌉. For example, what we would write as '$\forall p\ p$' in Leśniewski's notation is: '⌊p⌋⌈p⌉'.

1- Place Connectives

The basic constituent of each 1-place functor symbol is a horizontal line:'−'. By adding to this line vertical lines on its ends (two such lines are possible) in all possible combinations we obtain all possible 1-place functors. The rules are simple[16]:

- A vertical line occurs on the left side, iff (if the argument has the value 0, the whole expression has the value 1).
- A vertical line occurs on the right side, iff (if the argument has the value 1, the whole expression has the value 1).

Hence, we obtain four possible 1-place functors: ⊣, ⊢, ⊣⊢, −, which can be described by the following classical matrix:

---

[15] In (Leśniewski 1929) his notation seems to have been used for the first time, but, strictly speaking, Leśniewski explains it in his (1938a). It is commented on in Srzednicki and Stachniak (1988). (Luschei 1962, 289–305) also devotes some space to this *ideographic* notation.

[16] Leśniewski himself was not committed to the existence of truth-values, and he did not speak about them. He did not provide any formal semantics for his systems at all. I do say things in terms of truth-values but it does not entail that I am committed to their existence. For the sake of presentation I just take it to be a convenient abbreviation to speak about 'having truth value 0' instead of 'being false' and 'having truth value 1' instead of 'being true'.

p	H	−	⊣	⊢
1	1	0	1	0
0	1	0	0	1

The left side of the horizontal line corresponds to argument's having value 0 and the right side corresponds to argument's having value 1. A vertical stroke occurs if for that value the whole expression has value 1. 'H' is called *verum*: if the argument has value 0, the whole expression has value 1 (hence the stroke on the left), and if the argument has value 1, the whole expression has value 1 (hence the stroke on the right). '−' is called *falsum*. No matter whether the argument's value is 0 or 1, the value of the whole is 0 (hence no vertical strokes). '⊢' is negation. If the argument's value is 0, the value of the whole is 1 (hence the stroke on the left), and if the argument's value is 1, the whole has value 0 (hence no stroke on the right). '⊣' is called *assertion*. The value of the whole is 1 just in case the value of the argument is 1 (hence the stroke on the right and no stroke on the left).[17]

2- Place Connectives

The basis of a 2-place functor is a circle: 'o' (called 'the hub'). To this circle four possible lines ('spokes') can be added in four directions—up, down, left, right. The rules of adding these lines are as follows:

- A line to the left occurs iff for arguments having values respectively 1, 0, the whole has value 1.
- A line to the right occurs iff for arguments having values respectively 0, 1, the whole has value 1.
- A line upwards occurs iff for arguments of value 0, 0, the whole has value 1.
- A line downwards occurs iff for arguments of value 1, 1, whole has value 1.

These rules can be pictured by the following schema:

$$
\begin{array}{ccc}
 & 0\,0 & \\
 & | & \\
1\,0 \; — & \bigcirc & — \; 0\,1 \\
 & | & \\
 & 1\,1 &
\end{array}
$$

According to these rules we obtain 16 possible 2-place functors. To characterize them, we shall apply the following method: a 2-place functor will be characterized by a 4-place sequence of elements 1 and 0, simply by giving values taken by the whole expression when arguments have values, respectively, 11, 10, 01, 00.[18]

---

[17] (Luschei 1962, 292) calls verum the *tautologous connection of one argument* and falsum the *contradictory connection of one argument*.

[18] Names of functors are usually given according to Luschei's translation (Luschei 1962: 290–291), with slight differences.

Functor	Name	Characteristics
ϕ	Equivalence	1001
‑o‑	Disjunction	0110
ò	Conjunction	1000
‑ò‑	Exclusion	0111
ò	Conegation	0001
‑ϕ‑	2-place *verum*	1111
o	2-place *falsum*	0000
‑o	Distinction	0100
o‑	Contradistinction	0010
ϕ‑	Implication	1011
‑ϕ	Counterimplication	1101
‑ϕ‑	Alternation	1110
‑ϕ	Antecedent Affirmation	1100
ϕ‑	Consequent Affirmation	1010
ò‑	Antecedent Negation	0011
‑ò	Consequent Negation	0101

A peculiar feature of this notation is that there is a correspondence between geometrical properties of symbols of functors and the logical properties of these functors. I don't want to get into details (see, Urbaniak 2006), but here are two examples:

1. Let us define the relation of inclusion between two functors:

$$\mathbf{D} \subset: \ \lfloor fh \rfloor \lceil \phi \ (\lfloor rs \rfloor \lceil \phi \ (f(rs)h(rs))\rceil \subset (fh))\rceil$$

Notice first that in Leśniewski's notation the *definiens* comes first and the *definiendum* occurs on the right side of the formula. As we can see, we have defined the expression $\subset$ of category $\frac{s}{\frac{s}{ss}\frac{s}{ss}}$.

That is, we define a connective, $\subset$, which takes two 2-place sentential connectives as arguments and yields a sentential expression as a result. The variables f, h represent sentential connectives, and r, s are propositional variables. In modern notation:

$$\forall f, h\,[f \subset h \equiv \forall r, s\,(r\,f\,s \to r\,h\,s)]$$

The inclusion of 1-place functors is defined analogously (i.e. we should change the number of arguments of functors and the number of variables which the quantifier binds).

Now, it turns out that a functor h of the same number of arguments as f includes functor f in the above sense iff f is graphically a part (proper or not) of h. For example, ' ϕ ' geometrically includes the functor ' ò ', and it also includes it according to the above definition, because for any r, s it is true that if ò (r, s) then (r, s).

2. Define the functor cnv of category $\frac{s}{\frac{s}{ss}}$ by:[19]

---

[19] Observe that instead of using indices to mark the semantic category of a variable, Leśniewski used various shapes of brackets surrounding the arguments. The strategy, alas, is less systematic

$\mathsf{D}\,cnv\langle f\rangle: \quad \lfloor frs\rfloor\lceil\;\phi\;(f(sr)cnv\langle f\rangle(rs))\rceil$

which reads:

$$\forall f, r, s\,[cnv(f)(r, s) \equiv f(s, r)].$$

That is, applying cnv to a connective f yields a connective: $cnv(f)$ which applied to two sentential argument yields truth-value 1 iff f itself applied to those arguments but in reversed order yields truth value 1.

We may say that g is a converse of f iff g is equivalent to $cnv\langle f\rangle$. The geometrical rule is:

Two 2-place functors, say f, g are converses of each other iff both:

- f has a horizontal line on a given side iff g has a horizontal line on the opposite side, and
- f has a vertical line in a given direction iff g also has a vertical line in the same direction.

For example the following pairs are converses of each other:

$$\phi, \phi; \phi, \phi; \sigma, \sigma.$$

The notation can be systematically extended to connectives of more arguments while preserving the correlation between the shape and the logical relationship between functors (see Urbaniak, 2006).

3.4 Leśniewski's Protothetic and Principia Mathematica

A very important feature of Protothetic is that definitions are treated as expressions of this system, and equivalence (or implication, depending on the formulation) is treated as the only primitive connective. For this reason, Leśniewski's account of definitions varies from that of Russell's. Russell treated definitions as expressions of the meta–language used to introduce abbreviations. Leśniewski objected to this treatment on three counts:

1. If a definition is to be an expression of the meta-language, then the meaning of 'iff's in definitions is more vague than it would have been otherwise. It is not sure whether it contributes anything to the system, allowing to generate new non-equiform theorems, or whether it is rather just an unofficial way of abbreviating theorems of the basic language of the system. It is not obvious (at least, if you are as fastidious as Leśniewski was) whether such a definition is a rule of inference, a notational rule or just an expression of the author's state of mind.

---

than that of using indices because nowhere did he give a clear and general explanation of how the shape of brackets corresponds to the semantic category involved. In practice, this did not cause much confusion.

2. If having as simple a formulation as possible is an essential virtue of an axiomatic system (as Leśniewski believed), it would be useful to reduce definitions to something already present in the system.
3. As it turns out, some definitions in Leśniewski's systems are creative.[20] That is, introducing definitions may allow to prove something which does not contain the defined term, but is unprovable without definitions. But if definitions are to be treated only as meta-linguistic abbreviations, they should not have any "logical strength". In contrast, if a definition is just taken to be an axiom of a specific kind, the fact that its addition strengthens the system is less surprising. Some people believe that definitions should not be creative, some people don't. Leśniewski belonged to the latter group. (Also, see Chap. 6 for more details about Leśniewski's view of definitions.)

Leśniewski's criticism of Russell was actually a little bit more complex. Rumour has it that Leśniewski devoted one of his Warsaw seminars to *Principia Mathematica*.[21] It took him (the legend continues) one semester to read one page with his students. Indeed, a distinctive feature of Leśniewski's style is that he always very meticulously (and somewhat uncharitably) tries to reconstruct the view that he is about to criticize. To the best of my knowledge there is no written account of the content of the famous seminar. However, when Leśniewski started publishing his *On the Foundations of Mathematics* series, the first chapter (included in his 1927) titled "On some questions regarding the sense of the 'logistic' theses" was devoted to a criticism of *Principia Mathematica*.

Although Leśniewski had his first contact with something like modern formal logic already in 1911, it took him quite a few years to convince himself of the importance of symbolization for the development of logic and mathematics. The basic problem for him, it seems, was the lack of clarity as to the status of axioms and theorems of a given theory, to which the usual comments were no cure:

> The first encounter with 'symbolic logic' created within me a strong aversion to that discipline for a number of years to come. Even the exposition of its elements included in the book by Łukasiewicz, and certain other expositions of 'symbolic logic' which I attempted in succession, wishing somehow to absorb the results reached by the exponents of this science were incomprehensible to me, and not because of my own fault... Not possessing the faculty of entering into the spirit of other people's ideas, I was estranged from the science itself by the considerable effect of the obscure and ambiguous comments provided by its exponents. The decidedly sceptical dominant note of the position I occupied for a number of years in relation to 'symbolic logic' stemmed from the fact that I was not able to become conscious of the

---

[20] Słupecki (1955) contains a proof of this claim for Ontology, and (Slupecki 1953) mentions this property of Protothetic. Generally the idea is that one can think of the language of Protothetic or Ontology as a higher–order language. Leśniewski's formulation did not contain an axiom of comprehension. Instead, it contained the rule that every formula of a certain form defines a new constant. A version of the axiom of comprehension was obtainable by existential instantiation after the introduction of a constant. Hence, for instance, in the formulations of Ontology which contain certain form of the axiom of comprehension, rules of definition are no longer creative (stachniak, 1981).

[21] A story I heard from A.W. Mostowski in a discussion. I do not recall seeing the claim anywhere in print.

real 'sense' of the axioms and theorems of that theory—'of what' and 'what', respectively, it
was desired to 'assert' by means of the axioms and theorems. (Leśniewski, 1927, 181–182)

His criticism focused on the use of Russell's assertion sign ('⊢'), the meaning assigned to logical constants in *Principia Mathematica*, and Whitehead's and
Russell's comments about those symbols. He opens his discussion with the following words:

> The character of the semantical doubts which occupied me for some time during attempts,
> long futile, to read the works written by 'logicists' can easily be grasped by anybody by a
> careful analysis of the comments which Whitehead and Russell provide for particular types
> of expressions which occur in the 'theory of deduction', and he may also reflect on such
> an occasion how much the aforesaid comments are infused with refined cruelty towards a
> reader who is accustomed to paying some attention to anything he reads. (Leśniewski 1927,
> 182)

As an example, Leśniewski considers ⋆**1.3.**:

$$\vdash: q. \to .p \lor q$$

The comment one can find in *Principia Mathematica* is: "This principle states: if q
is true, then p or q is true". To figure out why this is the right reading Leśniewski
tries to find in *Principia Mathematica* an explanation of the meaning of expressions
like:

$$\vdash .p$$
$$p \lor q$$
$$p \to q$$

Turnstile

As for ⊢, two comments that Leśniewski finds in *Principia Mathematica* are:

> (3.1) "The sign '⊢', called the 'assertion-sign' means that what follows is asserted. It is
> required for distinguishing a complete proposition, which we assert, from any subordinate
> propositions contained in it but not asserted."

> (3.2) "In symbols, if 'p' is a proposition, 'p' by itself will stand for the unasserted proposition,
> while the asserted proposition will be designated by '⊢ .p'. The sign '⊢' is called the
> assertion-sign; it may be read 'it is true that' (although philosophically this is not exactly
> what it means)."

So, what is asserted by '⊢ .p'? Passage (3.1) suggests that the propositions that are
asserted are those that immediately follow the turnstile and a dot (so in this reading
it is p itself that is asserted). This reading is also supported by the following:

> (3.3) "The dots after the assertion-signs, indicate its range; that is to say, everything following
> is asserted, until we reach either an equal number of dots, preceding a sign of implication or
> the end of the sentence."

On the other hand, passage (3.2) tells us that it is the whole expression ⊢ .p that is asserted. This reading has also another textual support:

(3.4) "On all occasions where, in Principia Mathematica, we have an asserted proposition of the form '⊢ .fx' or '⊢ .fp' this is not to be taken as meaning '⊢ .$(x).fx$' or '⊢ .$(p).fp$'."

As Leśniewski puts it:

After reading all the passages of the work which refer to the assertion sign and expressions of the type '⊢ .p' the reader who is not particularly distinguished by exuberant frivolity can hardly say that he understands the intentions of the authors. (Leśniewski 1927, 183)

Indeed, Leśniewski emphasizes that so far the only clear indication as to how to read '⊢' in natural language is that one cannot read it as 'it is true that' without changing its meaning.

In his attempt to clarify, Leśniewski formulates three questions:

(1) Suppose 'p' is a proposition. Is '⊢ .p' also a proposition?

(2) Suppose 'p' is a proposition having a certain sense. Does '⊢ .p' have the same sense?

(3) What should we take to be Whitehead's and Russell's axioms and theorems? (a) whole expressions ('⊢ .' included) or (b) just those subexpressions which follow '⊢ .'?

Leśniewski then discusses a few initially plausible readings of '⊢' and their relation to the above questions.

Reading A. '⊢ .p' means 'we (Whitehead and Russell) assert that p'. The answer to question (1) is positive. The answer to question (2) is negative. The answer to question (3) is (a).

The problem here is that on this reading the axioms and theorems are very specific statements about the authors of the theory and express only the fact that the authors of the given theory assert this or that formula. As Leśniewski puts it:

...a system created from that type of proposition, is not a system of logic at all; It could rather be considered as a deductive confession of the authors of the given theory. (Leśniewski 1927, 187)

Reading B. '⊢ .p' means 'that which follows is asserted: p'. On this view, according to Leśniewski, the answer to question (1) is negative. For him,

that which follows, is asserted: p

is not a proposition, but rather "a peculiar composite of three parts following one after another, of which the first, the assertion sign, is a proposition in the form of a single expression having the same sense as the proposition "that which follows is asserted", the second is formed of dots, and the third, the expression 'p' is a proposition, as we assumed." (Leśniewski 1927, 185) Thus, the complex expression which is not a proposition does not have the same sense as its constituent proposition. The answer to question (3) on this reading is (b).

Leśniewski points out that if this is how one should read '⊢', it seems that White-head and Russell are not using the symbol consistently, for they sometimes put it in front of a formula which they certainly do not want to assert as theorems. For instance we can find in *Principia Mathematica* statements like:[22]

Similarly ⊢: $(y) : (\exists x).f(x, y)$.

Reading C. '⊢ .p' has the same meaning as 'p'. The authors put '⊢' in front of a proposition just to suggest that it is a proposition which they are asserting (but the assertion is not part of the content of a proposition). We should count as axioms whole expressions, '⊢' included. Leśniewski suggests that the same objection that he put forward against (B), *mutatis mutandis*, applies here.

Use of Variables and Semantic Ascent

Continuing his comments on ⋆**1.3.** Leśniewski notices that according to ⋆**1.01.** impli-cation '$p \to q$' is defined by '$\neg p \vee q$', where definitions are supposed to preserve the meaning. Thus he takes ⋆**1.3.** to mean the same as:

(3.5) '$\neg q . \vee .p \vee r$'

Now the question is how we should understand the negation. The explanation to be found in *Principia Mathematica* is:

If p is any proposition, the proposition 'not p' or 'p is false', will be represented by '$\neg p$'.

Thus it seems that according to Whitehead and Russell reading '$\neg p$' as either 'not p' or 'p is false' does not change the meaning of what is claimed. This is problematic for Leśniewski, because in 'p is false', the symbol p no longer represents a proposition the way it does in '$\neg p$'. When we say '...is false', the dots should be filled by a *name* of a sentence (or proposition), not by the sentence itself. For instance the expression 'Spiderman is bald is false' is not a well-formed sentence. Rather, it has to say something *about* the sentence, as in " 'Spiderman is bald' is false".[23] That being the case, the sentence:

'p' is false

does not mean the same as '$\neg p$'.[24] They may be equivalent, perhaps, but the former says something about a linguistic expression. The latter does not seem to do that. 'Not: Spiderman is bald' is not a sentence about a sentence: it is about Spiderman and his baldness (if you will), but it does not contain a name that refers to a sentence (and frankly, it doesn't sound like a correct sentence either).

Similarly, Whitehead and Russell explain the meaning of disjunction:

---

[22] Leśniewski refers to p. XXIV of the first volume of the 1910 edition of Principia Mathematica.

[23] Needless to say, the Spiderman example isn't originally Leśniewski's.

[24] I put aside the worries about variables occurring in quotations.

If p and q are any propositions, the proposition 'p or q', i.e. 'either p is true or q is true' where the alternatives are to be not mutually exclusive, will be represented by '$p \vee q$'.

But then (3.5) means the same as:

(3.6) ' 'q' is false' is true or 'p or r' is true.

which expands to:

(3.7) ' 'q' is false' is true or ' 'p' is true or 'r' is true' is true.

Again, 'or' occurs in the above, and since it is taken to mean the same as '$\vee$' we can easily climb the ladder of semantic ascent, allegedly preserving the meaning of the whole sentence:

(3.8) ' 'q' is false' is true or either ' 'p' is true' is true or ' 'r' is true' is true.

Quite correctly, however, Leśniewski points out that semantic ascent does not preserve the meaning. Propositions that mean the same are about the same. But, for instance, 'Paris is in France' is a sentence about Paris and its location, whereas ' 'Paris is in France' is true' is about a sentence and its relation to the world.

In General

Indeed, Whitehead and Russell were not perfectly clear or consistent in their use of '⊢' or propositional connectives. Nor were they very careful when they explained the classical connectives in terms of semantic ascent. This Leśniewski took to be quite a serious problem:

> ...the reader will constantly wilt under the weight of this medley of interpretations. The impenetrable chaos which is the actual result of the reader's effort aimed at grasping the sense of the 'primitive proposition' ⋆1.3., could be even intensified, if the reader refers to a series of other comments contained in the work of Whitehead and Russell and connected with the comments on the mentioned problems, which do not harmonize with the authors' comments already quoted or even with one another. (Leśniewski 1927, 194)

On one hand, Leśniewski's criticism does not seem as crippling as he himself (quite uncharitably) took it to be.[25] Many logicians had (and have) decent sense of understanding of what was going on in PM, despite those, after all minor, philosophical issues. On the other hand, the criticism led Leśniewski to ignore the assertion sign to and avoid any kind of treatment of complex formulas which takes them to be saying something about their constitutive formulas. This indeed was a sensible move.

---

[25] For instance, one interpretation is that Whitehead and Russell used '⊢' to precede those formulas which they took to belong to the system they were developing. Then '⊢ $p \vee \neg p$' could be read as: ' '$p \vee \neg p$' is a theorem of the system under consideration'. The answer to (1) is positive, the answer to (2) is negative, and the answer to (3) is (b). Perhaps, some occurrences of '⊢' which precede expressions which are not theorems of PM can be found in PM, but there is nothing wrong about just saying that indeed, Whitehead and Russell sometimes made mistakes and misused the symbol.

3.5 Axiomatic Formulations

What follows is a simplification of the original account. The main issue here is that Leśniewski was a nominalist and his *directives* which he used to describe the system were not meant to describe some abstract entity (say, a set of theorems). They rather described how a system-token seen as a bunch of inscriptions is to be developed.

Indeed, his view on logical systems is quite dynamic: theorems are written expressions, and the system physically grows or develops as new theorems and proofs are being written down. This specific feature was mirrored in Leśniewski's explanations. He was really concerned with speaking of his systems as if they were a bunch of inscriptions. So, for instance, depending on the stage of development of a system, some expressions may be meaningful or meaningless according to whether their definitions have been already written down. For the sake of simplicity of our description I will not follow this approach, instead presenting a somewhat 'Platonized' version of Protothetic using mainstream methods.

A few words about notational conventions. Variables p, q, r are propositional variables within the system, variables f, g are variables of categories different than s (the context will determine their category). ϕ, ψ, χ are meta–linguistic variables standing for formulas, ζ (possibly with numerical subscripts) are schematic meta–variables representing variables of various semantic categories and τ (possibly with numerical subscripts) is a schematic letter standing for constants (propositional constants or functors of other categories).[26]

System $\mathfrak{S}$

As discussed in (Leśniewski 1929, 423–438), it had two axioms:

$$(\mathfrak{S}.1)\ ((p \equiv r) \equiv (q \equiv p)) \equiv (r \equiv q)$$
$$(\mathfrak{S}.2)\ (p \equiv (q \equiv r)) \equiv ((p \equiv q) \equiv r)$$

and two rules:

[*Detachment for equivalence*] From $\psi \equiv \chi$ and ψ infer χ.

[*Substitution for equivalence*] From $\psi \equiv \chi$ and ϕ infer $\phi(\chi)$ (which results from ϕ by replacing some of occurrences of ψ in ϕ by χ).

The system constitutes a sound and complete axiomatization of the classical equivalential calculus. That is, given the language of the theory contains only propositional variables, brackets, and the equivalence symbols, the classical (semantical) consequence operation coincides with deducibility in this system.

$\mathfrak{S}$ is not functionally complete. That is, not all classical connectives can be defined in a language which contains classical equivalence but no quantification binding propositional variables.

---

[26] At one point, '(ζ)' is a name of a rule, but the context makes it clear and no ambiguities arise.

$\mathfrak{S}$ has an interesting feature. For any expression ϕ in the language of this system (that is, correctly constructed from propositional variables and the equivalence symbol) this expression is a theorem of $\mathfrak{S}$ iff any propositional variable in ϕ occurs an even number of times.[27]

System $\mathfrak{S}_1$

This system (Leśniewski 1929, 438–450) admitted quantifiers binding both propositional variables and first-order functor variables. The system is based on three axioms:

$$(\mathfrak{S}_1.1) \quad \forall p, q, r \, (((p \equiv r) \equiv (q \equiv p)) \equiv (r \equiv q))$$

$$(\mathfrak{S}_1.2) \quad \forall p, q, r \, ((p \equiv (q \equiv r)) \equiv ((p \equiv q) \equiv r))$$

$$(\mathfrak{S}_1.3) \, \forall g, p \, \{\forall f \, (g(p, p) \equiv [\forall r \, (f(r, r) \equiv g(p, p)) \equiv$$
$$\forall r \, (f(r, r) \equiv g(p \equiv \forall q \, q, p))] \equiv \forall q \, g(p, p)\}$$

Tarski (1923) has shown that the system is functionally complete. First, he defined conjunction by:

$$p \wedge q \equiv \forall f \, [p \equiv [\forall r \, (p \equiv f(r)) \equiv \forall r \, (q \equiv f(r))]]$$

Then, other standard connectives can be defined:[28]

$$\neg p \equiv (p \equiv \forall q \, q)$$
$$p \to q \equiv (p \equiv p \wedge q)$$
$$p \vee q \equiv (\neg p \to q)$$

Once these connectives are available, functional completeness straightforwardly follows.

---

[27] This seems to be the first form of the so-called Leśniewski-Mihailescu theorem (Mihailescu 1937). The difference is that Leśniewski proved the theorem for a language without negation. Mihailescu's version applied to a language with equivalence and negation. For such s language the theorem requires not only each variable but also the negation symbol to occur an even number of times.

[28] In fact, once the system enforces extensionality (which Tarski called *the law of substitution*):

$$\forall p, q, f \, [(p \equiv q) \wedge f(p) \to f(q)]$$

(or the corresponding inference rule), conjunction can be defined by means of a simpler formula:

$$p \wedge q \equiv \forall f \, (p \equiv (f(p) \equiv f(q)))$$

It is common in the literature to call quantifiers binding propositional variables only *first-order protothetical quantifiers*, and to call other quantifiers *higher-order protothetical quantifiers*. Interestingly, first-order propositional quantification alone would not suffice to define all truth–functional connectives using equivalence as the sole primitive connective (see Surma 1977 for details). It only allows one to define two propositional constants: false ($\forall p\, p$) and true ($\forall p\, (p \equiv p)$), and four functions of one argument: assertion ($ass(p) \equiv p$), negation ($\neg p \equiv [(\forall p\, p) \equiv p]$), the verum of p ($ver(p) \equiv (p \equiv p)$) and the falsum of p ($fal(p) \equiv (\neg p \equiv p)$) but this does not yield functional completeness. Higher-order protothetical quantification is needed to define conjunction.

Now, ($\mathfrak{S}_1$.3) might look convoluted, but once we employ definitions of other connectives, things get quite simple. First of all, $p \equiv \forall q\, q$ tells us that that p is equivalent to something which is false, so we can render the expression as $\neg p$. Secondly, the huge left-hand of the equivalence is just a fancy way of stating the conjunction of $g(p, p)$ and $g(\neg p, p)$ in terms of equivalence. Thus, the content of ($\mathfrak{S}_1$.3) boils down to:

$$\forall g, p\, [g(p, p) \wedge g(\neg p, p) \equiv \forall q g(q, p)]$$

which is a version of a *principle of bivalence*, because it intuitively speaking says that if something holds of a sentence and of its negation, it holds of all sentences.[29]

All systems under discussion, allow for conditional proofs: to prove a conditional, or its universal generalization, one can assume the antecedent and derive the consequent. Also, it is taken for granted that to prove a universal generalization of a formula it is enough to prove the formula itself. These rules were not listed explicitly, but observe that rules (α)-(γ) listed below do not contain any quantifier introduction rule.[30] And indeed, the moves in question were used without any additional explanation by Leśniewski and his students, starting already with Tarski (1923). Apart from that, $\mathfrak{S}_1$ employed six rules of inference, all of them originally named by a Greek letter:

(α) *Detachment for equivalence.*

(β) *Substitution*: this works almost like universal quantifier elimination in natural deduction systems. From $\forall \zeta\, \phi$ it allows one to infer the expression resulting from substituting in ϕ for all occurrences of ζ (free in ϕ) a simple expression of the language (a constant or a variable) given that no substituted variable becomes bound. The rule applies to propositional variables and functor variables as well.

(γ) *Distribution*: let all free variables in χ be $\zeta_1, \dots, \zeta_k$, and let all free variables in ψ be $\zeta_{k+1}, \dots, \zeta_m$. From $\forall \zeta_1, \dots, \zeta_m\, (\chi \equiv \psi)$ infer $\forall \zeta_1, \dots, \zeta_k\, \chi \equiv \forall \zeta_{k+1}, \dots, \zeta_m\, \psi$. An analogous rule pertains to universal quantifiers binding functor variables.

---

[29] Leśniewski, partially due to his criticism of Łukasiewicz and Kotrabiński's attempts to make sense of a third logical value, denied the philosophical plausibility of many-valued logics.

[30] "Leśniewski never published technical directives for his method, as Gentzen and Jaśkowski did for theirs in 1934-35, since he presented his intuitive deductions only as outlines demonstrating how to construct rigorous proofs according to his official directives." (Luschei 1962, 40)

(δ) *Definition (of a propositional constant):* $\tau_{\zeta_1,...,\zeta_k} \equiv \chi$ is a definition of a newly introduced propositional constant τ of category s with possible parameters $\zeta_1, \ldots, \zeta_k$ iff χ is a well-formed formula with $\zeta_1, \ldots, \zeta_k$ as the only free variables. If τ does not contain parameters, χ cannot contain any free variables.

(ε) *Definition (of a prototothetical functor):* Say $\mu_1, \ldots, \mu_v, \zeta_1, \ldots, \zeta_k$ are variables of categories $\sigma_1, \ldots, \sigma_v, \sigma_1', \ldots, \sigma_k'$. Then

$$\forall \mu_1, \ldots, \mu_v, \zeta_1, \ldots, \zeta_n (\tau_{\mu_1,\ldots,\mu_v}(\zeta_1, \ldots, \zeta_n) \equiv \chi(\mu_1, \ldots, \mu_v, \zeta_1, \ldots, \zeta_n))$$

is a definition of a newly introduced k-place functor τ of category $\frac{\sigma}{\sigma_1',\ldots,\sigma_k'}$ with parameters $\mu_1, \ldots, \mu_v$ iff $\mu_1, \ldots, \mu_v, \zeta_1, \ldots, \zeta_n$ are the only free variables in $\chi(\mu_1, \ldots, \mu_v, \zeta_1, \ldots, \zeta_k)$ (non-primitive expressions cannot occur in wffs unless they are earlier defined). An important aspect both of (δ) and of (ε) is that parameters do not contribute to the semantic category of the newly defined constant—we will see the importance of this fact in Sect. 3.6.

(ζ) A rule concerning quantifiers. Originally it was formulated for a language with implication as the primitive symbol. No clear description of this rule exists in the sources. It was supposed to move universal quantifier from the front of a conditional to the front of the consequent, given that no variable bound by this quantifier occurred as free in the antecedent. What we know is that Tarski in 1922 proved it to be dependent on the other rules and as such redundant (so we do not need it anyway). (Srzednicki and Stachniak 1988, 22–23)

System $\mathfrak{S}_2$

This system, described in (Leśniewski 1929, 451–53), built over a richer language than $\mathfrak{S}_1$ (now, both variables of all types constructible from s, and quantifiers binding them are allowed) is an extension of $\mathfrak{S}_1$ by a rule dubbed the *rule of verification*, also called (η). It originates from the following idea due to Łukasiewicz. Say two propositional constants are introduced:

$$0 \equiv \forall q \, q$$
$$1 \equiv (\forall q \, q \equiv \forall q \, q)$$

Take a formula ϕ whose all variables are propositional. Consider all possible substitutions of 0s and 1s for all propositional variables in ϕ. If they all are sentences provable in $\mathfrak{S}_1$, the universal generalization of ϕ is a theorem.

Leśniewski introduced an analogous rule for semantic categories other than s. For any semantic category σ there is a specific set of all possible extensional constants S (called *verifiers* of that category) such that for any formula ϕ if all results of substitution of free variables of category σ by elements of S are provable in $\mathfrak{S}_1$, then the formula ϕ preceded by universal quantifiers binding these variables is a theorem of $\mathfrak{S}_2$.

For instance, for the semantic category s, the verifiers are 1 and 0, as defined above. For $\frac{s}{s,s}$ verifiers are constituted by all 16 truth-functional connectives. In general the verifiers of a given semantic category are all (up to logical equivalence) differently defined constants of that category. Leśniewski provided a detailed explanation of how a complete list of definitions of verifiers of any specific category is to be produced (since this is not intrinsically complex, let us not elaborate on the details).

One thing worth observing is that even though this rule is quite intuitive, proofs employing higher-order quantifiers become quite lengthy. For instance, to prove $\forall f, p\,(\neg f(p) \vee f(p))$ one has to prove four other formulas:

$$\forall p\,(\neg\neg p \vee \neg p)$$
$$\forall p\,(\neg ass(p) \vee ass(p))$$
$$\forall p\,(\neg ver(p) \vee ver(p))$$
$$\forall p\,(\neg fal(p) \vee fal(p)).$$

This was one of the reasons why the rule was abandoned in later systems.

Note that Leśniewski did not want rule (η) to allow the introduction of universal quantifiers binding propositional variables. In his formulation, the rule (1) describes the definitions of all verifiers of a certain category other than s and (2) allows one to introduce $\forall \zeta\,\phi(\zeta)$ if all instances resulting from the substitution of verifiers of the same category as ζ for ζ in $\phi(\zeta)$ have been proven. A likely reason why this rule excluded the case of a universal quantifier binding a propositional variable is that this would make the rule partially redundant. Other rules handle quantifiers binding propositional variables sufficiently: it turns out that if an expression can be proved by applying (η) to propositional variables, then it is already provable in $\mathfrak{S}_1$.

System $\mathfrak{S}_3$

The system (Leśniewski 1929, 454–464) resulted from the fact that the formulation of (η) is somewhat lengthy and so are proofs employing this rule. Leśniewski proposed a rule of extensionality, called $(\eta\star)$ which turns out to do the same job as (η). The rule says that for any one-place functor of at least second order the corresponding law of extensionality (to be defined below) is a theorem. To formulate the law of extensionality for higher-order functors we have to go through a few definitions.

Definition 3.3 (*Order of a functor*) The order of a functor is defined by two conditions:

- Every functor all arguments of which belong to the category of propositions is a first-order functor.
- Every functor at least one argument of which is of order $k - 1$ and no argument is of order k or higher, is k-order. □

Definition 3.4 (*Higher-order equivalence*) 0-order equivalence is just material equivalence. First-order equivalence of sentential connectives is defined by:

$$\forall_{f,g,p_1,\dots,p_k}((f \doteq g) \equiv (f(p_1, \dots, p_k) \equiv g(p_1, \dots, p_k)))$$

Various kinds of equivalence between expressions of the same category σ are defined according to the following schema:

$$(\chi_1 \doteq^\sigma \chi_2) \equiv \forall \xi_1, \ldots, \xi_n \, (\chi_1(\xi_1, \ldots, \xi_n) \doteq^{\sigma'} \chi_2(\xi_1, \ldots, \xi_n))$$

where χ_1 and χ_2 belong to category $\sigma = \frac{\sigma'}{\sigma_1, \ldots, \sigma_n}$ and $\xi_1, \ldots, \xi_n$ are variables of categories (respectively) $\sigma_1, \ldots, \sigma_n$. That is, two functors of category $\sigma = \frac{\sigma'}{\sigma_1, \ldots, \sigma_n}$ are σ-equivalent iff applying them to the same arguments results in σ'-equivalent expressions.

Now, the rule of extensionality says that any instance of the following:

$$\forall \zeta_1, \zeta_2, \zeta \, [(\zeta_1 \doteq^\sigma \zeta_2) \to (\zeta(\zeta_1) \doteq^{\sigma'} \zeta(\zeta_2))]$$

is a theorem, where ζ_1, ζ_2 are variables of category σ, and ζ is a higher-order functor of category $\frac{\sigma'}{\sigma}$. This reads: if two expressions ζ_1 and ζ_2 are σ-equivalent, then for any one-place functor ζ of appropriate semantic category, the result of applying ζ to ζ_1 is σ'-equivalent to the result of applying it to ζ_2.

$\mathfrak{S}_3$ employs the same axioms as $\mathfrak{S}_1$ and the rules: (α), (β), (γ), (δ), (ε) and $(\eta\star)$. The system is equivalent to $\mathfrak{S}_2$, and is a sound and complete with respect to the classical semantics in which propositional variables range over two truth-values and variables of other categories range over extensional functions of appropriate type.

System $\mathfrak{S}_4$

This system, introduced in (Leśniewski 1929, 456–459)[31] is a conditional formulation of $\mathfrak{S}_3$ based on the following axioms:

$(\mathfrak{S}_4.1)\forall p, q \, (p \to (q \to p))$

$(\mathfrak{S}_4.2)\forall p, q, r \, ((p \to q) \to ((q \to r) \to (p \to r)))$

$(\mathfrak{S}_4.3)\forall p, q, r \, (((p \to q) \to r) \to ((p \to r) \to r))$

$(\mathfrak{S}_4.4)\forall g, p, q \, (g(p,q) \to (g(p \to \forall q \, q, p) \to g(q,p)))$

$(\mathfrak{S}_4.4)$ is quite interesting. Let us see what it says and why it is valid. It states that for any two–place connective g, if $g(p,q)$ is true and $g(p \to \forall q \, q, p)$ (that is, $g(\neg p, p)$, because $\forall q \, q$ is disprovable) is true, then also $g(q,p)$ is true. Why? Well, Protothetic is a two–valued extensional logic. Hence, under any valuation q has to be either (materially) equivalent to $\neg p$ or to p. If it has the same truth–value as $\neg p$, then we can substitute q for $\neg p$ in $g(\neg p, p)$ and hence $g(q, p)$. If, on the other hand, q has the same truth value as p, we can substitute p for q in $g(p,q)$ and so we get $g(p, p)$. But then, we also can substitute q for p and we get $g(q, p)$ as well.

---

[31] There, Leśniewski described his formulation from year 1922.

The rules of system $\mathfrak{S}_4$ are:

(α_I) *Detachment for implication*: From $\phi \to \psi$ and ϕ infer ψ.

(β_I) *Substitution*: just like (β), only for a language based on implication.

(γ_I) *Distribution*: This is an intuitive correlate of (γ). Let all free variables in χ be $\zeta_1, \ldots, \zeta_k$, and let all free variables in ψ be $\zeta_{k+1}, \ldots, \zeta_m$. From $\forall \zeta_1, \ldots, \zeta_m (\chi \to \psi)$ infer $\forall \zeta_1, \ldots, \zeta_k \chi \to \forall \zeta_{k+1}, \ldots, \zeta_m \psi$.

(δ_I) and (ϵ_I) *Rules of definitions*: they correspond to (δ) and (ϵ) and describe what definitions in a language based on implication should look like. The only new thing is that instead of material equivalence '$\phi \equiv \psi$' such definitions employ logically equivalent longer formulas built by means of material implication:

$$\forall \sigma (((\phi \to \psi) \to ((\psi \to \phi) \to \sigma)) \to \sigma)$$

where σ is a propositional variable which is free neither in ϕ nor in ψ.

(ζ_I) A conditional formulation of (ζ).

($\eta_I\star$) A conditional formulations of ($\eta\star$) (analogously, another variant of the system is obtained if instead of ($\eta\star$) a conditional version of (η) is employed).

Later (1922–1923), this system was simplified. The first important simplification is due to Tarski (see Lesniewski 1929, 459), who (1) suggested writing definitions in the form of two separate implications, and (2) showed that then it will suffice to use (($\mathfrak{S}_4$.1)) and the following *axiom of verification* instead of all the above axioms:

$$\forall f, p, q, r (f(r, p) \to (f(r, p \to \forall s \, s) \to f(r, q)))$$

System $\mathfrak{S}_5$

This is the final form of Protothetic. It results from a slight modification of $\mathfrak{S}_3$—the rule of definitions now puts *definienda* on the right side of the equivalence and not on the left. The difference is that this makes proofs slightly shorter, because it allows one to apply universal quantifier distribution and detachment to definitions without employing the theorem stating that equivalence is symmetric.

3.6 Some Theorems and Proofs

Now, let us take a look at some theorems and proofs in Protothetic.[32] First, to get the hang of the expressive power of the language of Protothetic, let us take a look at a few theorems of full Protothetic. We will focus on properties of connectives of category $\frac{s}{s}$ (analogous theorems for other semantic categories also exist).

---

[32] This section is based on (Słupecki 1953). My modifications consist in selecting the material to be included, adding natural language explanations, filling in some proof details, and re-arranging the proof layout.

$$\forall f, p, q \, ((p \equiv q) \rightarrow (f(p) \equiv f(q))) \tag{3.1}$$

(3.1) is the *law of extensionality* for 1-place connectives. It says that substitution of materially equivalent arguments in the scope of such connectives does not change the value of the whole expression.

$$\forall f \, (\forall p \, (f(p) \equiv ass(p)) \vee \forall_p(f(p) \equiv ver(p)) \vee \forall_p(f(p) \equiv fal(p)) \vee \forall_p(f(p) \equiv \neg p))$$
$$\tag{3.2}$$

(3.2) is the *law of the number of functions*. It says that a 1-place connective is equivalent to one of the four connectives defined before: $ass(p)$, $ver(p)$, $fal(p)$, $\neg p$.

$$\forall f, p \, (f(p) \equiv (f(1) \wedge p \vee f(0) \wedge \neg p)) \tag{3.3}$$

(3.3) is called the *law of development*. It says that any $f(p)$, when p is true, is equivalent to $f(1)$, and is equivalent to $f(0)$ otherwise.

$$\forall f, p \, (f(1) \wedge f(0) \rightarrow f(p)) \tag{3.4}$$

(3.4) is the *law of verification* for 1-place $\frac{s}{s}$ connectives. It says that if both $f(1)$ and $f(0)$ are true, so is $f(p)$ for any p.

As it turns out, none of (3.1–3.4) is $\mathfrak{S}_1$-provable, and all of them are pairwise $\mathfrak{S}_1$-equivalent—that is, their equivalence can be proven in Protothetic even without the use of the rule of verification and the rule of extensionality. Of particular importance is the $\mathfrak{S}_1$-equivalence of (3.1) and (3.4), because a generalization of the equivalence proof yields the result that verification and extensionality over the language of Protothetic have the same logical strength. This, of course, means that $\mathfrak{S}_2$ (based on the rule of verification) and $\mathfrak{S}_3$ (based on the rule of extensionality) are indeed equivalent, as indicated before.

Thus, to give a taste of what proofs in Protothetic look like, we will concentrate on proving the equivalence of the law of verification (3.4) and the law of extensionality (3.1). We will work within the implicational system $\mathfrak{S}_4$. Since we want to prove an $\mathfrak{S}_1$-equivalence, (η) and $(\eta\star)$ will not be used in these proofs.

One move that was practiced both by Leśniewski and by Leśniewskian scholars is the use of additional assumptions and conditional proofs (this is one reason to think that Leśniewski was one of the pioneers of natural deduction). Thus, if one wants to prove a conditional $\phi \rightarrow \psi$ or its universal generalization (the result of preceding it with a universal quantifier binding all free variables), it is enough to assume ϕ and derive ψ. We will allow ourselves to employ this strategy too.[33] Also, I will help myself to the mainstream method of writing natural deduction proofs Fitch-style. (The original style consisted rather in just listing the formulas and giving the formula numbers on which they depended.)

Recall the axioms of $\mathfrak{S}_4$. Let's see what we can prove from them.

---

[33] Natural deduction methods were in early stage of their development, so Leśniewski and his students helped themselves to some of those techniques without explicitly describing them.

Theorem 3.5 $(\mathfrak{S}_4.1)$, $(\mathfrak{S}_4.2)$, $(\mathfrak{S}_4.3) \vdash_{\mathfrak{S}_1} \forall p, q \, (((p \to q) \to p) \to p)$. *That is, formula:*

$$(\mathfrak{S}_4.3')\quad \forall p, q \, (((p \to q) \to p) \to p)$$

can be proved from these axioms using the rules available in $\mathfrak{S}_1$.

Proof

1	$(p \to q) \to p$	Assumption
2	$((p \to q) \to p) \to ((p \to p) \to p)$	Sub p/r: $(\mathfrak{S}_4.3)$
3	$(p \to p) \to p$	MP: 1, 2
4	p	Assumption
5	$p \to p$	Discharge: $4 \Rightarrow 4$
6	p	MP: 3, 5

To prove a generalized conditional we assume its antecedent in line 1. Line two is obtained by substituting p for r in $(\mathfrak{S}_4.3)$. Detachment (also known as *Modus Ponens*, hence MP) applied to line 1 and 2 yields line 3. To prove $p \to p$ it is enough to assume p in line 4, observe that it directly follows from itself and discharge this assumption in line 5. Then, MP applied to lines 3 and 5 gives us p, which ends the proof.

The reason why $(\mathfrak{S}_4.3')$ is interesting is that once universal quantifiers are dropped, $(\mathfrak{S}_4.1)$, $(\mathfrak{S}_4.2)$ and $(\mathfrak{S}_4.3')$ together give us the Tarski-Bernays axiomatization of the purely implicational classical calculus (as its name indicates, the axiomatization was formulated by Łukasiewicz (1929). Theorem 3.5 is good news because since we know that Tarski-Bernays axiomatization is complete for the purely implicational calculus, so is the purely implicational part of $(\mathfrak{S}_1)$ (by the way, since we know the Tarski-Bernays axiomatization is complete, we also know that $(\mathfrak{S}_4.1)$, $(\mathfrak{S}_4.2)$, $((\mathfrak{S}_4.3')) \vdash_{\mathfrak{S}_1} (\mathfrak{S}_4.3)$. This also means that henceforth we will be able to rely on $(\mathfrak{S}_4.3')$ (in general, in proofs we can use previously proven theorems).

The next question is what happens once we take the universal generalization of the implicational axioms and add the Tarski-style definition of negation ($\neg p$ means $p \to \forall p \, p$). Can we derive all classical propositional theorems that can be formulated using negation and implication? To fix ideas, take an axiomatization of the full classical calculus (again, Łukasiewicz's):

$$[Ł1]\quad \forall p \, ((\neg p \to p) \to p)$$
$$[Ł2]\quad \forall p, q \, (p \to (\neg p \to q))$$

Theorem 3.6 *Both claims hold:*

$$(\mathfrak{S}_4.1), (\mathfrak{S}_4.2), (\mathfrak{S}_4.3) \vdash_{\mathfrak{S}_1} [Ł1]$$
$$(\mathfrak{S}_4.1), (\mathfrak{S}_4.2), (\mathfrak{S}_4.3)] \vdash_{(\mathfrak{S}_1)} [Ł2]$$

Proof For the first claim, we will use $(\mathfrak{S}_4.3')$:

$$
\begin{array}{ll}
1 & ((p \to \forall p\, p) \to p) \to p \qquad \text{Sub } \forall p\, p/q : (\mathfrak{S}S_4.3') \\
2 & (\neg p \to p) \to p \qquad\qquad\qquad \text{Def } \neg: 1
\end{array}
$$

Line one is obtained from $(\mathfrak{S}_4.3')$ by substituting $\forall p\, p$ for q. Line 2 just follows by definition of negation. The fact that I substituted a complex expression in line 1 might seem worrying, given that Leśniewski originally only allowed for substitution of simple expressions. This, however, is not a problem, because substitution of complex expressions is handled in Protothetic by the rule of substitution together with the rules of definitions. Thus, instead of substitution $\forall p\, p$, I could first define $\bot$ as $\forall p\, p$, then substitute $\bot$ for q in $(\mathfrak{S}_4.3')$, and then use the definition of $\bot$ to derive line 1. Since this only makes proofs longer without providing any more insight into their structure, from now on I will apply substitution to complex expressions without apology.

For the second claim, consider the following proof:

$$
\begin{array}{lll}
1 & p & \text{Assumption} \\
2 & p \to \forall p\, p & \text{Assumption} \\
3 & \forall p\, p & \text{MP: } 1, 2 \\
4 & q & \text{Sub } q/p: 3 \\
5 & (p \to \forall p\, p) \to q & \text{Discharge: } 2 \Rightarrow 4 \\
6 & p \to ((p \to \forall p\, p) \to q) & \text{Discharge: } 1 \Rightarrow 5 \\
7 & p \to (\neg p \to q) & \text{Def} \neg: 6
\end{array}
$$

To prove line 7, by definition, it is enough to prove line 6. So we assume its antecedent in line 1 and aim at proving its consequent in line 5. But that consequent is an implication itself. So we assume its antecedent in line 2 and try to prove its consequent in line 4. MP applied to 1 and 2 gives us line 3, and plain substitution applied to line 3 gives us the desired result in line 4.

This tells us that we have (the universal generalization of) the whole classical logic at our disposal. From now on I will just apply a generic rule called CL to mean that something follows by the classical logic, without explaining how this is to be proven using the axioms and primitive rules of the system.

Now, our goal is a proof of the equivalence of extensionality and verification for 1-place sentence connectives. That is, we will be trying to prove:

[VerEx] $\forall f, p, q\, ((p \equiv q) \to (f(p) \equiv f(q))) \equiv \forall f, p\, (f(1) \wedge f(0) \to f(p))$

Theorem 3.7 $(\mathfrak{S}_4.1), (\mathfrak{S}_4.2), (\mathfrak{S}_4.2) \vdash_{\mathfrak{S}_1} [VerEx]$.

Proof $\Rightarrow$: We will assume extensionality, assume $f(1)$ and $f(0)$, and derive $f(q)$ without assuming anything about q.

1	$\forall f, p, q\, [(p \equiv q) \rightarrow (f(p) \equiv f(q))]$	Assumption
2	$f(1) \wedge f(0)$	Assumption
3	$q \equiv 1 \vee q \equiv 0$	CL
4	$(q \equiv 1) \rightarrow (f(q) \equiv f(1))$	Sub $q/p, 1/q$: 1
5	$(q \equiv 0) \rightarrow (f(q) \equiv f(0))$	Sub $q/p, 0/q$: 1
6	$(f(q) \equiv f(1)) \vee (f(q) \equiv f(0))$	CL: 3, 4, 5
7	$f(q)$	CL: 2, 6
8	$\forall q\, (f(1) \wedge f(0) \rightarrow f(q))$	Discharge: 2$\Rightarrow$7

$\Leftarrow$: Now it's time for a constant with a parameter to come to the rescue. We will define a constant Φ_p which contains a variable in its subscript, and yet is a one-place constant forming expressions like $\Phi_p(q)$:

$$[Def_{\Phi_p}]\ \forall p, q\, [\Phi_p(q) \equiv \forall f\, ((p \equiv q) \rightarrow (f(p) \equiv f(q)))]$$

From now on I will not only help myself to (universal generalizations of) classical propositional theorems, but also to standard classical treatment of quantifiers (any such move will be marked by 'CL$\forall$'). Now, we will need a few claims:

Lemma 3.1 *All of the following are $\mathfrak{S}_1$-provable from $(\mathfrak{S}_4.1)$, $(\mathfrak{S}_4.2)$ and $(\mathfrak{S}_4.3)$:*

(a)	$\Phi_0(1)$
(b)	$\Phi_0(0)$
(c)	$\Phi_1(1)$
(d)	$\Phi_1(0)$
(e) $\forall p, q\, (\Phi_p(q) \equiv \Phi_q(q))$	

For (a) consider the following proof:

1	$\neg(0 \equiv 1)$	CL
2	$\forall f\, ((0 \equiv 1) \rightarrow (f(0) \equiv f(1)))$	CL$\forall$: 1
3	$\Phi_0(1)$	Def$_\Phi$: 2

For (b) take:

$$
\begin{array}{ll}
1 & \forall p, q\,((p \equiv p) \to (q \equiv q)) \qquad \text{CL} \\
2 & 0 \equiv 0 \to (f(0) \equiv f(0)) \qquad \text{Sub } 0/p,\, f(0)/q \colon 1 \\
3 & \Phi_0(0) \qquad\qquad\qquad\qquad\qquad \text{Def}_\Phi \colon 2
\end{array}
$$

Proofs of (c) and (d) are similar.
(e) holds because equivalence is symmetric:

$$
\begin{array}{lll}
1 & \Phi_p(q) & \text{Assumption} \\
2 & \forall f\,((p \equiv q) \to (f(p) \equiv f(q))) & \text{Def}_\Phi \colon 1 \\
3 & \quad q \equiv p & \text{Assumption} \\
4 & \quad p \equiv q & \text{CL} \colon 3 \\
5 & \quad f(p) \equiv f(q) & \text{CL}\forall \colon 2, 4 \\
6 & \quad f(q) \equiv f(p) & \text{CL} \colon 5 \\
7 & \forall f\,((q \equiv p) \to (f(q) \equiv f(p))) & \text{Discharge: } 3 \Rightarrow 6 \\
8 & \Phi_q(p) & \text{Def}_\Phi \colon 7
\end{array}
$$

Proof in the other direction in analogous.

Now we can give the desired proof, deducing extensionality from verification:

$$
\begin{array}{lll}
1 & \forall f, p\,(f(1) \wedge f(0) \to f(p)) & \text{Assumption} \\
2 & \Phi_1(1) \wedge \Phi_1(0) \to \Phi_1(p) & \text{Sub } \Phi_1/f \colon 1 \\
3 & \Phi_0(1) \wedge \Phi_0(0) \to \Phi_0(p) & \text{Sub } \Phi_0/f \colon 1 \\
4 & \Phi_1(p) & \text{CL: (c), (d), 2} \\
5 & \Phi_0(p) & \text{CL: (a), (b), 3} \\
6 & \Phi_p(1) & \text{CL}\forall \colon \text{(e), 4} \\
7 & \Phi_p(0) & \text{CL}\forall \colon \text{(e), 5} \\
8 & \Phi_p(1) \wedge \Phi_p(0) \to \Phi_p(q) & \text{Sub } \Phi_p/f \colon 1 \\
9 & \forall p, q\, \Phi_p(q) & \text{CL}\forall \colon 6, 7, 8 \\
10 & \forall f, p, q\,((p \equiv q) \to (f(p) \equiv f(q))) & \text{Def}_\Phi \colon 9
\end{array}
$$

The key moves are in lines 2, 3 and 8, where we substitute a defined constant with a parameter for a one-place variable. This is the first way definitions in Leśniewski's systems can help prove theorems. In this case, a definition allowed us to prove something about a relation between p and q by means of a claim about 1-place connectives.

This ends our short tour around protothetical proofs. One important thing to notice is that the rule of extensionality is formulated for one-place functors. The trick we used above can be generalized to prove that together with the rules of definitions, the rule of extensionality is strong enough to prove extensionality for n-ary functors of arbitrary semantic category, for any finite n. There is also another way definitions in Leśniewski's systems provide proofs with extra power, but its discussion will have to wait till Chap. 6.

3.7 Further Readings

Historically speaking, Protothetic was preceded by certain suggestions concerning propositional quantification. Russel (1906) suggested introducing quantifiers binding the propositional variables, but he did not develop this idea. Łukasiewicz (1929) developed a calculus where quantifiers binding propositional variables play an important part (see also Łukasiewicz and Tarski 1930), but it seems that the generality that Protothetic can afford was Leśniewski's idea. The very first paper related to Protothetic which employs quantification not only over propositional variables but also over variables representing connectives is Tarski (1923, reprinted in Srzednicki and Stachniak 1998), where it was shown that when this sort of generality is admitted, conjunction can be defined in terms of equivalence as the only connective. Tarski's results on functional completeness of Protothetic were extended by Sobociński (1949, reprinted in Srzednicki and Stachniak 1998), who generalized Tarski's considerations concerning forms of definitions in Protothetic. Sobociński's paper is also interesting for historical reasons—in the introduction he describes the fate of *Collectanea Logica*, the Polish journal in which (Leśniewski (1938a,b) were to appear. The outbreak of war made its publication impossible.

The project of finding the shortest possible axiomatization of Protothetic, already initiated by Leśniewski, was continued not only by him but also by some of his students for quite a few years. (Sobociński 1954, 1960, 1961) gives good surveys of the results. The latter paper also contains an interesting metatheorem which states a sufficient condition for an axiomatization of Protothetic (although no proof of its sufficiency is given). Sobociński also gave another, slightly more complicated version of sufficient conditions for completeness, also without a proof (but with a convincing explanation). Le Blanc (1985) provides another, shorter version of the axiomatization.

One of the best introductions to Protothetic so far is Słupecki (1953). The formulations that Słupecki focuses on are mainly conditional. His explanations of the rules of inference are somewhat brief, but he provides many examples of simple proofs which allow the reader to get an idea as to what rules he is using. In one aspect his rules differ from those of Leśniewski: he allows for substitution of complex expressions in place of variables of appropriate semantic categories, which was forbidden in the original formulation (this move, however, is conservative). A nice feature of Słupecki's presentation is that he starts with relatively simple systems and goes on to construct full Protothetic by gradual strengthening. Slupecki also discussed some interesting theorems that can be formulated in the language elementary Protothetic, which are also provably equivalent in elementary Protothetic (for one-place function symbols, at least, but a generalization of his results is to be found by Chikawa (1967)). A sketch of a completeness proof is included.

Rickey (1972, 1973), provides a good insight into the nature of Leśniewski's terminological explanations, with special attention to how the inscriptional description of the rules of Protothetic is to be understood.

An interesting book (in French) intended as an introduction to Protothetic is (Miéville 2001).

Leśniewski's idiosyncratic *wheel-and-spoke* notation is interesting for at least two reasons: (1) symbols for propositional one- and two-place connectives encode their truth-tables with their own shape, (2) geometric relations between different symbols correspond to various logical relations between the functions that they express. Recently it has been shown that there is a simple algorithm that extends this method to arbitrary first-order classical connectives (Urbaniak 2006).

A comprehensive monograph on intuitionistic Protothetic is (López-Escobar and Miraglia 2002).

Another group of formulations of Protothetic, hinted at by Leśniewski, is called *computative Protothetic* and was the subject of a PhD thesis at the University of Manchester (Le Blanc 1991). In the system there are signs for assertion and rejection, and rules describe which formulas are asserted or rejected on the basis of what has been previously asserted or rejected. The oldest system of computative Protothetic and contains nine rules of inference and is based on one axiom: '$A : \Lambda \rightarrow \Lambda$' (where Λ is the falsum constant and A means assertion). Le Blanc describes sixteen equivalent formulations, based on ten different combinations of primitive terms. He reconstructs those systems and provides a detailed description of their rules. Also, sketches of consistency, completeness and equivalence proofs are given.

An interesting research direction is many-valued Protothetics. Those have been studied (and a completeness result has been given) by Scharle (1971). It turns out that finitely-valued Protothetic can be axiomatized with the following property: for every closed wff ϕ one of $J_i(\phi)$ is provable, where J_i is a function that takes one of the designated values only if its argument has value i. The full proof of this fact can be given using a technique developed by Surma (1976) which uses finitely generated trees. Many-valued Protothetics have been the subject of a PhD dissertation at the University of Manchester, supervised by Lejewski (Watanabe 1973). As Watanabe has shown, the change in the number of admissible values results in the possibility of introducing different general quantifiers of one and the same semantic category, corresponding to different conjunctions one can introduce in a many-valued system.

There is a connection between Protothetic and Henkin's theory of propositional types (Henkin 1950, 1963), which was an improvement of the system of Church (1940). (Andrews (1963) constitutes an improvement of Henkin's theory). Minor details aside, Henkin's theory of propositional types can be thought of as a version of Protothetic.

Grzegorczyk (1964) shows that every one of Henkin's equality symbols is definable by mean of relational abstraction and the equality symbol for the usual material equivalence (although still the set of axioms is denumerable). Meridth (1951) provides also a similar system, commented on by Arai and Tanaka (1966).

Propositional calculi become more complex when we admit non-extensional operators. One of the intuitive steps is to introduce an identity operator which expresses something like 'expresses the same proposition' or 'states the same fact'. If the criteria of identity are more fine-grained than in Frege's account (according to which the denotation of any two sentences having the same truth value is the same), we get non-extensional calculi. Zero-order calculi with additional axioms and rules for identity, called Sentential Calculi with Identity (SCI), have been extensively studied by Suszko (1966, 1967, 1971a,b,c,d,e,f, 1972a,b, 1973, 1975, 1977) and developed by Omyła (1994). Apparently quite independently, following a suggestion of Prior (1957), Cresswell constructed a second-order theory of intensional identity. Besides the difference of level, in Suszko's system identity is a primitive symbol introduced axiomatically, whereas in Cresswell's system identity is defined by: '$p = q \equiv \forall f\,(f(p) \equiv f(q))$' where functor variables are allowed to range over non-extensional functors as well. The question of the relation between Cresswell's and Suszko's systems has not been fully explored. Also, Suszko at one point suggests that Protothetic would have been much more interesting if it had admitted the intensional identity. No such extension of Protothetic has been constructed so far. Another fascinating issue seems to be the relation between intensional identity and modalities. In his papers Suszko also shows how in some systems with identity as the only primitive non-extensional system it is possible to obtain various modal calculi.

References

Ajdukiewicz, K. (1935). Die syntaktische Konnexität. *Studia Philosophica, 1*, 1–27.

Andrews, P. (1963). A reduction of the axioms for the theory of propositional types. *Fundamenta Mathematicae, 52*, 345–350.

Arai, Y., & Tanaka, S. (1966). A remark on propositional calculi with variable functors. *Proceedings of the Japan Academy, 42*, 1056–1057.

Chikawa, K. (1967). On equivalences of laws in elementary prototetics I, II. *Proceedings of the Japan Academy, 43, 44*, 743–747, 56–59.

Church, A. (1940). A formulation of the simple theory of types. *Journal of Symbolic Logic, 5*, 56–68.

Grzegorczyk, A. (1964). A note on the theory of propositional types. *Fundamenta Mathematicae, 54*, 27–29.

Henkin, L. (1950). Completeness in the theory of types. *Journal of Symbolic Logic, 15*, 81–91.

Henkin, L. (1963). A theory of propositional types. *Fundamenta Mathematicae, 52*, 323–334.

Kotarbiński, T. (1929). *Elementy Teorii Poznania, Logiki Formalnej i Metodologii Nauk [Elements of the theory of knowledge, formal logic and methodology of the sciences]*. Lwów: Ossolineum.

Le Blanc, A. (1985). Investigations in protothetic. *Notre Dame Journal of Formal Logic*, *20*, 483–489. Included in Srzednicki and Stachniak (1998).

Le Blanc, A. (1991). Leśniewski's computative protothetic. Ph.D. Thesis, University of Mancherster.

Leśniewski, S. (1916). Podstawy ogólnej teoryi mnogości I. Prace Polskiego Koła Naukowego w Moskwie, 2 [Foundations of the general theory of sets I (Leśniewski 1991, pp. 129–173)].

Leśniewski, S. (1927). O Podstawach Matematyki, Wstęp. Rozdział I: O pewnych kwestjach, dotyczących sensu tez 'logistycznych'. Rozdział II: O 'antynomji' p. Russella, dotyczącej 'klasy klas, nie będących własnemi elementami'. Rozdział III: O różnych sposobach rozumienia wyrazów 'klasa' i 'zbiór'. Przegląd Filozoficzny, 30, 164–206 [On the foundations of mathematics. Introduction. Ch. I. On some questions regarding the sense of the 'logistic' theses. Ch. II. On Russel's 'antinomy' concerning 'the class of classes which are not elements of themselves'. Ch. III. On various ways of understanding the expression 'class' and 'collection' (Leśniewski 1991, 174–226)].

Leśniewski, S. (1938a). Einleitende Bemerkungen zur Fortsetzung meiner Mitteilung u.d.T. 'Grundzüge eines neuen Systems der Grundlagen der Mathematik'. Widener Library Info Harvard Depository XLL 270.5, Hollis number: 005913328 [Introductory remarks to the continuation of my article: 'Grundzüge eines neuen Systems der Grundlagen der Mathematic' (Leśniewski 1991, pp. 649–710)].

Leśniewski, S. (1938b). Grundzüge eines neuen Systems der Grundlagen der Mathematik, §12. WidenerLibrary Info Harvard Depository XLL 270.6, Hollis number: 002222243.

Leśniewski, S. (1928). O podstawach matematyki, Rozdział IV: O podstawach ogólnej teoryj mnogości I. *Przegląd Filozoficzny*, *31*, 261–291. On the foundations of mathematics. Ch. IV On 'Foundations if the general theory of sets. I' (Leśniewski 1991, pp. 227–263).

Leśniewski, S. (1929). Grundzüge eines neuen systems der grundlagen der mathematik §1-11. *Fundamenta Mathematicae*, *14*, 1–81. Fundamentals of a new system of the foundation of mathematics, §1-11 (Leśniewski 1991, pp. 410–605).

Leśniewski, S. (1930a). O podstawach matematyki, Rozdział VI: Aksjomatyka 'ogólnej teorji mnogości', pochodząca z r. 1918. Rozdział VII: Aksjomatyka 'ogólnej teorji mnogości', pochodząca z r. 1920. Rozdział VIII: O pewnych ustalonych przez pp. Kuratowskiego i Tarskiego warunkach, wystarczających i koniecznych do tego, by p było klasąp-tów a. Rozdział IX: Dalsze twierdzenia 'ogólnej teorji mnogości', pochodzące z lat 1921–1923. *Przegląd Filozoficzny*, *33*, 77–105. On the foundations of mathematics. Ch. VI. The axiomatization of the 'general theory of sets' fro the year 1918. Ch. VII. The axiomatization of the 'general theory of sets' from the year 1920. Ch. VIII. On certain conditions established by Kuratowski and Tarski which are sufficient and necessary for P to be the class of objects A. Ch. IX. Further theorems of the 'general theory of sets' from the years 1921–1923 (Leśniewski 1991, pp. 315–349).

Leśniewski, S. (1930b). Über die Grundlagen der Ontologie. *Sprawozdania z posiedzeń Towarzystwa Naukowego Warszawskiego, Wydział Nauk Matematyczno-Fizycznych*, *23*, 111–132. On the foundations of Ontology (Leśniewski 1991 pp. 606–628).

Leśniewski, S. (1931a). O podstawach matematyki, Rozdział X: Aksjomatyka 'ogólnej teorji mnogości pochodząca z r. 1921. Rozdział XI: O zdaniach 'jednostkowych' typu '$A\varepsilon b$. Przegląd Filozoficzny, 34, 142–170. On the foundations of mathematics. Ch. X. The axiomatization of the 'general theory of sets' from the year 1921. Ch. XI. On 'singular' propositions of the tyle '$A\varepsilon b$ (Leśniewski 1991 pp. 350–382).

Leśniewski, S. (1931b). Über definitionen in der sogenannten theorie der deduction. *Sprawozdania z posiedzeń Towarzystwa Naukowego Warszawskiego, Wydział Nauk Matematyczno-Fizycznych*, *24*, 289–309. On definitions in the so-called theory of deduction (Leśniewski 1991 pp. 629–648).

López-Escobar, E., & Miraglia, F. (2002). *Definitions: The primitive concept of logics or the Leśniewski-Tarski Legacy, Dissertationes Mathematicae* (Vol. 401). Warszawa: Polska Akademia Nauk.

Łukasiewicz, J. (1929). Elementy logiki matematycznej. [Reprinted in 1958 in Warsaw by Państwowe Wydawnictwo Naukowe and in 2008 in Poznań, Wydawnictwo Naukowe UAM].

Łukasiewicz, J. (1921). Logika dwuwartościowa. *Przegląd Filozoficzny*, *23*, 189–205.

Łukasiewicz, J., & Tarski, A. (1930). Untersuchungen über den Aussagenkalkül. *Comptes rendus de la Société des Sciences et des Lettres de Varsovie, 3*(23), 1–21.

Luschei, E. (1962). *The logical systems of Leśniewski*. Amsterdam: North-Holland.

Meredith, C. (1951). On an extended system of the propositional calculus. *Proceedings of the Royal Irish Academy, 54A,* 37–47.

Miéville, D. (2001). *Introduction à l'oeuvre de S. Lesniewski. Fascicule I: La protothétique*. Travaux de logique. Université de Neuchâtel.

Mihailescu, E. (1937). Recherches sur un sous-systéme du calcul des propositions. *Annales Scientifiques de l'Université de Jassy, 23,* 106–124.

Omyła, M. (1994). Zasady niefregowskiej semantyki zdań [Principles of non-fregean semantics of propositions]. In J. Pelc & L. Koj (Eds.), *Znaczenie i Prawda, Rozprawy Semiotyczne* (pp. 247–260). Warsaw: Naukowe PWN.

Prior, A. (1957). *Time and modality*. Oxford: Oxford University Press.

Rickey, F. (1972). Axiomatic inscriptional syntax I. *Notre Dame Journal of Formal Logic, 13,* 1–33.

Rickey, F. (1973). Axiomatic inscriptional syntax—part II: the syntax of protothetic. *Notre Dame Journal of Formal Logic, 14,* 1–52. Included in Srzednicki and Stachniak (1998).

Russell, B. (1906). The theory of implication. *American Journal of Mathematics, 28,* 158–202.

Scharle, T. (1971). Completeness of many-valued protothetic. *Journal of Symbolic Logic, 36,* 363–364. Abstract.

Słupecki, J. (1953). St. Leśniewski's protothetic. *Studia Logica, 1,* 44–112. Included in Srzednicki and Stachniak (1998).

Słupecki, J. (1955). St. Leśniewski's calculus of names. *Studia Logica, 3,* 7–72.

Sobociński, B. (1949). An investigation of protothetic. *Cahiers de l'Institut d'Études Polonaises en Belgique, 5,* 1–44.

Sobociński, B. (1954). Z badań nad aksjomatykąPrototetyki Stanisława Leśniewskiego [From investigations on the axiomatics of Stanisław Leśniewski's Protothetic]. *Rocznik Polskiego Towarzystwa Naukowego na Obczyźnie, 4,* 18–20.

Sobociński, B. (1960). On the single axioms of protothetic I. *Notre Dame Journal of Formal Logic, 1,* 52–73. Included in Srzednicki Stachniak (1998).

Sobociński, B. (1961). On the single axioms of protothetic II. *Notre Dame Journal of Formal Logic, 2,* 111–126, 1929–2148. Included in Srzednicki and Stachniak (1998).

Srzednicki, J., & Stachniak, Z. (Eds.). (1988). *S. Leśniewski's lecture Notes in logic*. Dordrecht: Kluwer Academic.

Srzednicki, J., & Stachniak, Z. (Eds.). (1998). *Leśniewski's Systems Stachniak*. Dordrecht: Kluwer Academic.

Stachniak, Z. (1981). *Introduction to model theory for Lesniewski's ontology*. Wrocław: Uniwersýtetu Wrocławskiego.

Stone, M. (1937). Note on formal logic. *American Journal of Mathematics, 59,* 506–514.

Surma, S. (1976). An algorithm for axiomatizing every finite logic. In D. Rine (Ed.), *Computer Science and Multiple-Valued Logic* (pp. 137–143). North-Holland.

Surma, S. (1977). On the work and influence of Stanisław Leśniewski. In R. Gandy & J.M.E. Hyland (Ed.), Logic Lolloquium 76, Studies in Logic and the Foundations of Mathematics, pp. 191–220. Amsterdam: North-Holland.

Suszko, R. (1966). Non-creativity and translatability in terms of intentions. *Logique et Analyse, 9,* 360–363.

Suszko, R. (1967). An essay in the formal theory of extension and of intension. *Studia Logica, 20,* 7–36.

Suszko, R. (1971a). Identity connective and modality. *Studia Logica, 27,* 7–39.

Suszko, R. (1971b). Quasi-completeness in non-Fregean logic. *Studia Logica, 29,* 7–16.

Suszko, R. (1971c). Reifikacja sytuacji [Reification of situations]. *Studia Filozoficzne, 2,* 65–82.

Suszko, R. (1971d). Semantics for the sentential calculus with identity. *Studia Logica, 28,* 77–82.

Suszko, R. (1971e). Sentential calculus with identity (SCI) and G-theories. *Journal of Symbolic Logic, 36,* 709–710.

Suszko, R. (1971f). Systemy S4 i S5 Lewisa a spójnik identyczności [Lewis's S4 and S5 and the identity connective]. *Studia Logica, 29,* 169–181.

Suszko, R. (1972a). Investigations into sentential calculus with identity. *Notre Dame Journal of Formal Logic, 13,* 289–308.

Suszko, R. (1972b). SCI and modal systems. *Journal of Symbolic Logic, 37,* 436–437.

Suszko, R. (1973). Adequate models for the non-Fregean sentential calculus (SCI). In Logic, language and probability. A selection of papers of the 4th Congress for Logic, Methodology and, Philosophy (pp. 48–54).

Suszko, R. (1975). Abolition of the fregean axiom. *Lecture Notes in Mathematics, 453,* 169–239.

Suszko, R. (1977). The fregean axiom and polish mathematical logic in the 1920s. *Studia Logica, 36,* 376–380.

Tarski, A. (1923). O wyrazie pierwotnym logistyki. *Przegląd Filozoficzny [Philosophical Movement], 26,* 68–69. Translated as "On the primitive term of logistic" in Logic, semantics, metamathematics, Clarendon Press, Oxford 1956, 1–24.

Urbaniak, R. (2006). On representing sentential connectives of Leśniewski's elementary protothetic. *Journal of Logic and Computation, 16*(4), 451–460.

Watanabe, S. (1973). On many-valued protothetics. PhD thesis, University of Manchester.

Chapter 4
Leśniewski's Ontology

Abstract After describing basic intuitions underlying the system, I briefly point out how it differs from the classical logic. (The main difference is that there is no distinction between predicates and singular terms: there is only one class of name variables, and names can refer to any number of objects.) Then, I describe its language and its original axiomatization. I give some examples of definitions in Ontology to highlight the flexibility of its language. Next, I discuss Leśniewski's argument against universals formulated in the language of Ontology, the relation between Ontology and Russell's description theory and Leśniewski's attempt to apply Ontology to a paradox concerned with four-dimensional objects. Secondary literature pertaining to Ontology is surveyed in the last section.

4.1 Ontology: Basic Intuitions

The first remark on Ontology, a logical system developed by Leśniewski, dates back to his introduction to "On the Foundations of Mathematics". Ontology is a second layer of his systems. The first layer, Protothetic, was concerned with propositional aspects of reasoning, widely understood. The second layer extends this by introducing a category of names (and other derivative categories) and a predication copula: ε. Leśniewski says that Ontology…

> …forms a certain kind of modernized 'traditional logic', and which in content and 'power' most nearly approaches the Schröder 'Klassenkalkül', considered as including the theory of 'individuals'. (Leśniewski 1927, 176)

The first axiomatization of Ontology[1] Leśniewski claims to have formulated while teaching a course 'Exercises in the domain of the Cantorian Theory of Sets' at Warsaw University in 1919/1920 (Leśniewski 1931, 367). The first official presentation

---

[1] All other formulations of Ontology consist in replacing the axiom of Ontology with another, provably equivalent axiom (or axioms).

R. Urbaniak, *Leśniewski's Systems of Logic and Foundations of Mathematics*,
Trends in Logic 37, DOI: 10.1007/978-3-319-00482-2_4,
© Springer International Publishing Switzerland 2014

of Ontology was Leśniewski's lecture in Warsaw on January 10, 1921, titled 'On
the principles of Ontology' at the scientific session of the Polish Association for
Psychology (Leśniewski 1931, 369).

Although Ontology is mostly concerned with predication, it differs from the classical predicate logic in a few respects. First of all, it introduces terms, but it makes
no distinction between names and predicates. In this respect, it follows traditional
syllogistics rather than the predicate calculus. In Ontology there is only one category
of names, no matter whether they are empty, singular or common terms. So, a wide
understanding of the word *name* is accepted here. A name is any expression that
may take the place of either M or N in a sentence of the form: M is N. Or, in other
words, all countable *categorematic* (as opposed to *syncategorematic*) expressions
that are not sentences are taken to be names. Some examples of names in this wide
understanding are: *a cat, Socrates, brown, the bald guy in the corner, unicorn*. Thus,
by including the case when a name does not refer, Leśniewski's logic is one of the
first formalized free logics.[2] The main motivation for this move is the same as standard motivations for free logics: one wants logic to be applicable to our discourse
independently of often empirical question whether the names used refer to anything
(and whether they refer to one or more objects).

Leśniewski's Ontology can be viewed as an extension of name calculus in the
wide sense of the word 'name' mentioned above. 'Name calculus' because the basic
specific variables $(a, b, c, d, \dots)$ that are used in Ontology are name variables (this
does not mean that no variables of other categories are used). Roughly speaking, an
intuitive way to think of name variables is that they correspond to possible countable
noun phrases or that they represent ways in which a name can refer. 'An extension'
because the language also contains variables of higher categories, insofar as those
categories can be "built" by combining categories of names and of sentences.

Another respect in which Ontology differs from standard first-order logic is that
it represents predication in a different way than first order languages. Besides quantifiers[3] and classical propositional connectives, Ontology has another primitive symbol, the epsilon connective ε (read as 'is' or 'is one of'). An expression of the form
$a\varepsilon b$ is read as 'a is b' (or 'a is one of b's') and it is true if and only if a is taken to be
a singular term that names an object which is among the objects named by b. So, the
following are true: *Socrates ε philosopher, The book on my desk ε Kleene's 'Introduction to Metamathematics'*, and the following turn out to be false (although still
well-formed): '*Unicorn ε animal*', '*elephant ε animal*'. Some examples of the rendering of simple quantified statements: 'some philosophers are mad' ('some b is c')
has the form $\exists a\,(a\,\varepsilon\,b \wedge a\,\varepsilon\,c)$ and 'all mad people think they are philosophers' ('all
b are c') becomes something like $\forall a\,(a\,\varepsilon\,b \rightarrow a\,\varepsilon\,c).$ [4]

---

[2] A free logic is a logic which holds also in empty domains. A free logic is supposed to avoid the
ontological commitment of other systems. For instance, standard first order logic requires that the
domain should contain at least one object.

[3] Historically speaking, Leśniewski used only the universal quantifier.

[4] However, it is possible that natural language quantification is ambiguous. So, for example, a
universal statement with existential import should be rather rendered as '$\forall a\,(a\,\varepsilon\,b \rightarrow a\,\varepsilon\,c) \wedge
\exists c\,c\,\varepsilon\,a$'.

Third, in Ontology quantifiers are not meant to carry ontological commitment. It is the use of ε by which existence is stated. So, existence is defined:

$$\forall a \, (ex(a) \equiv \exists b \, b\varepsilon a) \tag{4.1}$$

and it is provable in Ontology that for some a, a does not exist:

$$\exists a \, \neg ex(a) \tag{4.2}$$

This and similar facts raised a fairly complex debate about the interpretation of Leśniewski's quantifiers and the meaning of variables in his systems. For it is not clear how they are to be interpreted given Leśniewski's nominalism and the fact that he did not provide a semantics for his system. For now, I will ignore these problems, but I will list related literature in Sect. 4.9 and discuss the issue in Chap. 8.

4.2 The Language of Ontology

The notion of *semantic category* as applied to Ontology is pretty much the same as the notion which we already know from Sect. 3.2. What is new in comparison to Prototethic is that expressions of the category of names (n) are introduced, and expressions of categories that can be built from s's and n's are allowed, pretty much as in Prototethic we allowed categories built from s's only (see Definition 3.1 on p. 60).

All (atomic) formulas of type $a \, \varepsilon \, b$ are of category s. All terms are of category n. Any expression which with k arguments of categories $\sigma_1, \ldots, \sigma_n$ yields an expression of category σ is of category $\frac{\sigma}{\sigma_1 \cdots \sigma_n}$. For instance, ε takes two arguments, both of category n and forms an expression of category s. Thus, its category is $\frac{s}{nn}$.

The language of Ontology arises from a generalization parallel to that of Prototethic: for any semantic category of Ontology, constants (possibly with parameters) can be defined, and variables together with quantifiers binding those variables can be introduced.

Definition 4.1 Ontology is constructed in a language determined by the following[5]:

- a, b, c, d, e, possibly with numerical subscripts are nominal variables, of category n.[6]

---

[5] This is not exactly the language that Leśniewski used, but (granted the slightly Platonic approach) it has all the properties essential for the further development. Section 6.4 is devoted to a brief explanation of the specific aspects of Leśniewski's meta-logic.

[6] Leśniewski used unofficial notation in which a name variable in a formula was written in capital letter if the satisfaction of that formula required the uniqueness of the referent for that variable. Since I rather prefer a streamlined syntax which does not depend on model-theoretic properties of formulas, I will not follow Leśniewski on this.

- The language of Protothetic is a sub-language of the language of Ontology. That is, any formation rule that holds for Protothetic, hold also for Ontology and every well–formed formula of Protothetic is a well–formed formula of Ontology.
- ε is a primitive symbol of category $\frac{s}{nn}$.
- For any defined constant (possibly with parameters) of category $\frac{\iota}{\sigma_1 \cdots \sigma_n}$ (rules of definitions are given below),[7] there also are variables of that category, $f^{\overline{\frac{\tau}{\sigma_1 \cdots \sigma_n}}}$, $g^{\overline{\frac{\tau}{\sigma_1 \cdots \sigma_n}}}$, possibly with numerical subscripts. In practice, the superscripts are omitted when the semantic category is obvious from the context.[8]
- If ϕ is of category $\frac{\tau}{\sigma_1 \cdots \sigma_n}$ and expressions $\alpha_1, \ldots, \alpha_n$ are of categories $\sigma_1, \ldots, \sigma_n$, then $\phi(\alpha_1, \ldots, \alpha_n)$ is a well-formed expression of the language of Ontology of category τ.
- If ϕ is an expression of category s and ζ a variable of any category, then $\forall \zeta \, \phi$ is a well-formed formula of category s.   □

4.3 The Axiomatic System

Ontology, in its version from 1920 as described by Leśniewski[9] in his(Leśniewski 1931, 364–369), has a single axiom:[10]

$$\forall a, b \, [a \, \varepsilon \, b \equiv \exists c \, (c \, \varepsilon \, a) \land \forall c, d \, (c \, \varepsilon \, a \land d \, \varepsilon \, a \to c \, \varepsilon \, d) \land \forall c \, (c \, \varepsilon \, a \to c \, \varepsilon \, b)]$$

This axiom says that for any a and b, a is b iff (1) for some c, c is a, (2) for any c and d, if c is a and d is a, then c is d, and (3) for any c, if c is a,then c is also b.

The intended interpretation of ε[11] is that $a \, \varepsilon \, b$ is true if and only if there is exactly one object which is a, and this object is among the objects which are b (this includes the case where b names exactly one object). Accordingly, (1) is supposed to guarantee the non-emptiness of a, (2) is intended to express the requirement that a names at most one object, and (3) is meant to state the inclusion between a and b ('any object which is a, is b').

---

[7] Leśniewski did not have a specified list of constants. Rather, he invented symbols ad hoc as he went. Sometimes he took symbols from the language of set theory which had similar meaning, sometimes he took abbreviations of Latin words, sometimes he used Greek capital letters, but there does not seem to be any deep uniformity in his choice of constant symbols.

[8] This is not Leśniewski's original device. His main tool was introducing different shapes of brackets surrounding the arguments.

[9] There are various other axiomatizations, see Sect. 4.9 for references.

[10] However, it has certain axiomatic rules that allow for introduction of an infinite number of theorems without any premises.

[11] Which is in no way determined by the axiomatic basis of Ontology, see Urbaniak (2006b, 2009) for details.

In a somewhat streamlined version (equivalent to the original) one can use standard rules for quantifiers and apply propositional logic just as is it done in predicate logic.[12] That is, the quantifier elimination and introduction procedures are just an extension of classical rules to variables of higher categories. Besides there are a few specific rules, especially connected with definitions and introduction of variables of different semantic categories.

Any definition of a constant name expression of category n (possibly with parameters $\zeta_1, \ldots, \zeta_k$) has the form:

$$\forall a \, [a \, \varepsilon \, \Delta_{\zeta_1,\ldots,\zeta_k} \equiv a \, \varepsilon \, a \wedge \phi(\zeta_1, \ldots, \zeta_k)]$$

where $\Delta_{\zeta_1,\ldots,\zeta_k}$ is the defined term and ϕ is a formula with a and $\zeta_1, \ldots, \zeta_k$ as its only free variables.

Protothetical definitions: for any variables $\alpha_1, \ldots, \alpha_k$ of semantic categories $\sigma_1, \ldots, \sigma_k$ and any formula β with $\alpha_1, \ldots, \alpha_k$ (and possibly $\zeta_1, \ldots, \zeta_u$) as its only free variables, there are constants τ, τ' of categories $\frac{s}{\sigma_1\cdots\sigma_k}$ and $\frac{n}{\sigma_1\cdots\sigma_k}$ (possibly with parameters $\zeta_1, \ldots, \zeta_u$), the former introduced by the universal closure of (4.3), the latter introduced by the universal closure of (4.4).

$$\tau_{\zeta_1,\ldots,\zeta_u}(\alpha_1, \ldots, \alpha_k) \equiv \beta \tag{4.3}$$

$$a \, \varepsilon \, \tau'_{\zeta_1,\ldots,\zeta_u}(\alpha_1, \ldots, \alpha_k) \equiv a \, \varepsilon \, a \wedge \beta \tag{4.4}$$

Many-linked definitions: say

$$\chi_1^{\sigma_1} \, \cdots \, \chi_k^{\sigma_1} \text{ are variables of category } \sigma_1$$
$$\chi_1^{\sigma_2} \, \cdots \, \chi_l^{\sigma_2} \text{ are variables of category } \sigma_2$$
$$\vdots \quad \vdots \quad \vdots \qquad\qquad \vdots \qquad\qquad \vdots$$
$$\chi_1^{\sigma_m} \, \cdots \, \chi_n^{\sigma_m} \text{ are variables of category } \sigma_m$$

and β is a formula with the above listed variables (and possibly $\zeta_1, \ldots, \zeta_u$) as its only free variables. Then there is a constant ν of category:

$$\frac{s}{\underbrace{\sigma_1, \ldots, \sigma_1}_{k}, \underbrace{\sigma_2, \ldots, \sigma_2}_{l}, \ldots, \underbrace{\sigma_m, \ldots, \sigma_m}_{n}}$$

defined by the universal closure of:

$$\nu_{\zeta_1,\ldots,\zeta_u}(\chi_1^{\sigma_1}, \ldots, \chi_k^{\sigma_1})(\chi_1^{\sigma_2}, \ldots, \chi_l^{\sigma_2}) \cdots (\chi_1^{\sigma_m}, \ldots, \chi_n^{\sigma_m}) \equiv \beta$$

---

[12] Originally, the rules for quantifiers and propositional inference were a trivial extension (to quantification over variables of different categories) of the rules employed in Protothetic, see Chap. 3 for details.

Finally, the (two-fold) rule of extensionality is:

- If α and β are nominal variables and γ is a nominal variable distinct from them both, any formula of the form:

$$\forall \gamma \, [\gamma \, \varepsilon \, \alpha \equiv \gamma \, \varepsilon \, \beta] \to \forall \chi \, [\chi(\alpha) \equiv \chi(\beta)]$$

is a theorem (where χ is a variable of category $\frac{s}{n}$).
- Further, suppose:

  - τ^1 and τ^2 are constants of category $\sigma = \frac{s}{\sigma_1 \cdots \sigma_k}$, (possibly with parameters $\zeta_1, \ldots, \zeta_u$),
  - $\tau^{1'}$ and $\tau^{2'}$ are constants of category $\sigma' = \frac{n}{\sigma_1 \cdots \sigma_k}$, (possibly with parameters $\zeta_1, \ldots, \zeta_u$),
  - $\alpha_1, \ldots, \alpha_k$ are distinct variables of categories $\sigma_1, \ldots, \sigma_k$,
  - ζ_s is a variable of category $\frac{s}{\sigma}$,
  - ζ_n is a variable of category $\frac{s}{\sigma'}$,

  Then all instances of the following two schemata are theorems:

$$\forall \zeta_1, \ldots, \zeta_u, \alpha_1, \ldots, \alpha_k \, [\tau^1_{\zeta_1, \ldots, \zeta_u}(\alpha_1, \ldots, \alpha_k) \equiv \tau^2_{\zeta_1, \ldots, \zeta_u}(\alpha_1, \ldots, \alpha_k)] \to$$
$$\to \forall \zeta_s, \zeta_1, \ldots, \zeta_u \, [\zeta_s(\tau^1_{\zeta_1, \ldots, \zeta_u}) \equiv \zeta_s(\tau^2_{\zeta_1, \ldots, \zeta_u})]$$

$$\forall \zeta_1, \ldots, \zeta_u, \alpha_1, \ldots, \alpha_k \, [a \, \varepsilon \, \tau^{1'}_{\zeta_1, \ldots, \zeta_u}(\alpha_1, \ldots, \alpha_k) \equiv a \, \varepsilon \, \tau^{2'}_{\zeta_1, \ldots, \zeta_u}(\alpha_1, \ldots, \alpha_k)] \to$$
$$\to \forall \zeta_n, \zeta_1, \ldots, \zeta_u \, [\zeta_n(\tau^{1'}_{\zeta_1, \ldots, \zeta_u}) \equiv \zeta_n(\tau^{2'}_{\zeta_1, \ldots, \zeta_u})]$$

Roughly speaking, the rule of extensionality says that if two names refer to the same things, or two functors apply to the same arguments, then whatever is true of one of them, is also true of the other.

4.4 Some Examples of Definitions in Ontology

To appreciate the flexibility of the language of Ontology, let us take a look at some examples of definitions that can be built within its language. Coextensiveness is defined by:

$$\forall a, b \, (\circ(a, b) \equiv \forall c \, (c \, \varepsilon \, a \equiv c \varepsilon b))$$

Sometimes it is simpler to use the infix notation (e.g. to write '$a \circ b$' instead of '$\circ(a, b)$'). Obviously, $\circ$ is[13] symmetric, reflexive and transitive:

---

[13] In the context where I describe theorems of Ontology (or theorems of Mereology) I will just say 'it is true' or 'holds', still meaning 'it is provable from the axioms of the system under consideration using its rules'.

$$\forall a, b \, (a \circ b \equiv b \circ a)$$

$$\forall a \, a \circ a$$

$$\forall a, b, c \, (a \circ b \wedge b \circ c \rightarrow a \circ c)$$

The existence of a's ('there is at least one a') is defined by:

$$\forall a \, (ex(a) \equiv \exists b \, b \, \varepsilon \, a)$$

'ex' is of semantic category $\frac{s}{n}$. We can also introduce variables and quantification for that category. Using the rule of extensionality we can easily prove that[14]:

$$\forall a, b \, (a \circ b \rightarrow \forall f^{\frac{s}{n}} \, (f^{\frac{s}{n}}(a) \equiv f^{\frac{s}{n}}(b))) \tag{4.5}$$

We can also define inclusion between two expressions of category $\frac{s}{n}$:

$$\forall f, g \, [f \subseteq g \equiv \forall a \, (f(a) \rightarrow g(a))] \tag{4.6}$$

Reflexivity and transitivity for $\subseteq$ thus defined hold.

Boolean operations on expressions (union, intersection, difference) of category $\frac{s}{n}$ can be defined as well.

$$\forall a, f, g \, [f \cup g(a) \equiv f(a) \vee g(b)]$$

$$\forall a, f, g \, [f \cap g(a) \equiv f(a) \wedge g(b)]$$

$$\forall a, f, g \, [f - g(a) \equiv f(a) \wedge \neg g(a)]$$

(Note that in the above we are dealing with complex functors '$f \cup g$', '$f \cap g$', '$f - g$' applied to a single argument a, not with the union of f and $g(a)$ etc.)

We can 'climb' up the ladder of semantic categories. For instance, we can define non-emptiness of a $\frac{s}{n}$-expression, in the way analogous to that used for defining existence.

$$\forall f \, [ex(f) \equiv \exists a \, f(a)]$$

The functor $ex^{\frac{s}{\frac{s}{n}}}$ defined this way is an example of an expression of the category $\frac{s}{\frac{s}{n}}$.

We can define generalized Boolean operations on expressions of category $\frac{s}{n}$. For instance, the intersection of all names of which f is truly predicable can be defined by:

$$\forall f, a \, [a \, \varepsilon \, \cap (f) \equiv \forall b \, (f(b) \rightarrow a \, \varepsilon \, b)]$$

---

[14] Basically because $a \circ b$ means $\forall c \, (c \, \varepsilon \, a \equiv c \, \varepsilon \, b)$ so (4.5) results from an instance of the rule of extensionality pretty much by definition of $\circ$.

Similar definitions are available for other operations and other categories. For instance, ∘ for two expressions of category $\frac{s}{nn}$ would be defined[15]:

$$\forall f, g \, (f \circ g \equiv \forall a, b \, (f(a, b) \equiv g(a, b))$$

For a functor f of category $\frac{s}{nn}$ we can easily define the domain $d(f)$ and range $r(f)$ of it:

$$\forall f, a \, [d(f)(a) \equiv \exists b \, f(a, b)]$$
$$\forall f, a \, [r(f)(a) \equiv \exists b \, f(b, a)]$$

For any $\frac{s}{nn}$-expression, its restriction to specific domains and ranges is introduced by:

$$\forall f, g, h, a, b \, [restr(f, g, h)(a, b) \equiv f(a, b) \wedge g(a) \wedge h(b)]$$

For any two-place functor we can define its reflexivity, transitivity, symmetry, anti-symmetry. Thus we can also express the fact that it is a partial or well-ordering. Pretty much all the standard notions can be introduced: first, least, maximal, greatest element of a given relation, least upper bounds, greatest lower bounds, chains etc. (with respect to a restricted domain or not).

4.5 Against Universals, Revisited

As an example of how the language of Ontology may be used to formalize philosophical arguments I will now describe one of Leśniewski's arguments against the existence of universals.

(Leśniewski (1927), p. 199) constructed an argument against the existence of universals, very similar to the one from (Leśniewski (1913), pp. 50–53).[16] The difference is that it does not employ the notion of property and it does not employ negative properties. In its original version it consists of one assumption and six conclusions following subsequently. Even though it is not formalized in Ontology, the way it is phrased makes the formalization relatively easy. Here is Leśniewski's original version. Just like his early argument against universals, this one starts with a very specific assumption about abstract objects:

---

[15] Mind that in our official notation the full form of this definition should be

$$\forall f^{\frac{s}{n,n}}, g^{\frac{s}{n,n}} \, (\circ^{\frac{s}{\frac{s}{n,n} \cdot \frac{s}{n,n}}} (f^{\frac{s}{n,n}}, g^{\frac{s}{n,n}}) \equiv \forall a, b \, (f^{\frac{s}{n,n}}(a, b) \equiv g^{\frac{s}{n,n}}(a, b)))$$

[16] The latter has been already discussed in Sect. 2.9.

However, in connection with that passage and with reference to all those who, by reason of
the meaning they give to expressions of the type 'general object'[17] with respect to objects
a,[18] are inclined to state the proposition 'if X is a general object with respect to objects a, X
is b and Y is a, then Y is b',[19] I wish to state here that this proposition entails the proposition
'if there exist at least two different a, then a general object with respect to objects a does
not exist,' in accordance with the following schema:

(1) If X is a general object with respect to objects a, X is b, [and] also Y is a, then Y is b.
    (assumption)

(2) If X is a general object with respect to objects a, X is different from Z, and Z is a, then Z
    is different from Z.

(3) If X is a general object with respect to objects a, X is identical with Z and Y is a, then Y
    is identical with Z.

    From (2) it follows that:

(4) If X is a general object with respect to objects a, and Z is a, then X is identical with Z.

(5) If X is a general object with respect to objects a, Z is a and Y is a, then (X is a general
    object with respect to objects a, X is identical with Z, and Y is a).

    From (5) and (3) it follows that:

(6) If X is a general object with respect to objects a, Z is a and Y is a, then Y is identical with
    Z.

    From (6) however it follows that if there exist at least two different a, then a general object
    with respect to objects a does not exist.

Formalized reconstructions of the full argument have been given by Waragai
(1981) and Gryganiec (2000). I will sketch a simplified proof which starts with the
formalization of Leśniewski's assumption and ends with his final conclusion.

Let us start with the assumption that if c is a general object with respect to objects
a and c is b, then any object d which is a is also b. The intuition here is pretty much
the same as the one behind Leśniewski's first argument. We start with a bunch of
objects, all named by a name a, and formulate a necessary condition for an object, say
c, to be a general object with respect to those objects. Namely, whatever is predicable
of this general object should be predicable about any of the objects among a's.

If we represent *is* using the epsilon, and treat '*gen*' as a one-place functor which
with one name argument, say a, constructs a name $gen(a)$ ('The general objects with
respect to a's'), this assumption may be formalized by:

$$\forall a, b, c, d\, [c \,\varepsilon\, gen(a) \rightarrow (c \,\varepsilon\, b \rightarrow (d \,\varepsilon\, a \rightarrow d \,\varepsilon\, b))] \qquad (4.7)$$

---

[17] By 'general object' Leśniewski understood universals.

[18] For instance, in this sense (at first approximation), redness would be a general object with respect
to all red objects.

[19] Leśniewski informally used capital letters as name variables in formulas if the truth of a formula
required that only one object be assigned to a variable. So, instead of a, if the truth of a formula
requires that only one object is a, Leśniewski would use a capital letter, like X, Y or Z. For the sake
of uniformity I do not use capital letters as name variables in the formalism.

The intended conclusion is that if c is a general object with respect to objects a, d is a, and e is a, then d is identical with e, that is, there are no objects which are general with respect to more than one object. We can formalize the claim as follows:

$$\forall a, c, d, e \, [c \, \varepsilon \, gen(a) \wedge d \, \varepsilon \, a \wedge e \, \varepsilon \, a \rightarrow d \, \varepsilon \, e \wedge e \, \varepsilon \, d] \tag{4.8}$$

To give the proof of (4.8) from (4.7) we need two easy theorems of Ontology:

$$\forall a, b \, (a \, \varepsilon \, b \rightarrow a \, \varepsilon \, a) \tag{4.9}$$

$$\forall a, b, c \, (a \, \varepsilon \, c \wedge b \, \varepsilon \, c \wedge c \, \varepsilon \, c \rightarrow a \, \varepsilon \, b \wedge b \, \varepsilon \, a) \tag{4.10}$$

I skip their proofs. A few words why the above holds, though. Observe that $c \, \varepsilon \, c$ guarantees that exactly one object is c. But if both the object which is a and the object which is b are c, they have to be one and the same object. Now, let us prove (4.8) from (4.7):

1	$c \, \varepsilon \, gen(a)$	Assumption
2	$d \, \varepsilon \, a$	Assumption
3	$e \, \varepsilon \, a$	Assumption
4	$c \, \varepsilon \, c$	CL$\forall$: 1, (4.9)
5	$c \, \varepsilon \, gen(a) \rightarrow (c \, \varepsilon \, c \rightarrow (d \, \varepsilon \, a \rightarrow d \, \varepsilon \, c))$	Sub: (4.7)
6	$c \, \varepsilon \, gen(a) \rightarrow (c \, \varepsilon \, c \rightarrow (e \, \varepsilon \, a \rightarrow e \, \varepsilon \, c))$	Sub: (4.7)
7	$c \, \varepsilon \, c \rightarrow (d \, \varepsilon \, a \rightarrow d \, \varepsilon \, c)$	MP: 1, 5
8	$c \, \varepsilon \, c \rightarrow (e \, \varepsilon \, a \rightarrow e \, \varepsilon \, c)$	MP: 1, 6
9	$d \, \varepsilon \, a \rightarrow d \, \varepsilon \, c$	MP: 4, 7
10	$e \, \varepsilon \, a \rightarrow e \, \varepsilon \, c$	MP: 4, 8
11	$d \, \varepsilon \, c$	MP: 2, 9
12	$e \, \varepsilon \, c$	MP: 3, 10
13	$d \, \varepsilon \, c \wedge e \, \varepsilon \, c \wedge c \, \varepsilon \, c \rightarrow d \, \varepsilon \, e \wedge e \, \varepsilon \, d$	Sub: (4.10)
14	$d \, \varepsilon \, e \wedge e \, \varepsilon \, d$	CL: 13, 11, 12 and 4

Leśniewski's comments:

I regard my treatment as the result of a careful formulation of theoretical tendencies involved, more or less explicitly, in the argumentation of opponents of the different kinds of 'universals' in various phases of their 'disputes' about them. If one takes the position that this assertion is a banal one, I would cite in defence the circumstance that exponents of 'philosophy' defend, regrettably often, positions at variance with banal assertions. (1927, 199)

The philosophical importance of the formal result is debatable. It may be equally well taken to prove that if there are general objects (i.e. universals), they do not satisfy (4.7). Indeed, a sensible proponent of universals is quite likely to deny that if an object falls under a universal, it has all the properties that the universal does (for instance, no concrete object is abstract even though any universal that it falls under is).

4.6 Description Theory and Ontology

Recall that one of the basic theorems about Russell's definite descriptions states:

$$A(\iota x(B(x))) \equiv \exists y\, [\forall z\, (B(z) \equiv z = y) \wedge A(y)] \qquad (4.11)$$

(The only x which is B is A iff there is a y such that any z is B iff it is y, and y is A.)

In Ontology, instead of things like 'A(x)' we rather have to say '$b\,\varepsilon\,a$', where x corresponds to b and A to a. Thus, we can recast (4.11) as[20]:

$$\forall a, b\, [b\,\varepsilon\,a \equiv \exists c\, [\forall d\, (d\,\varepsilon\,b \equiv d = c) \wedge c\,\varepsilon\,a]] \qquad (4.12)$$

As it turns out (Hiż 1977), (4.12) is equivalent to one of the (many equivalent) axioms of Ontology:

$$\forall a, b\, [b\,\varepsilon\,a \equiv \exists c\, c\,\varepsilon\,b \wedge \forall c\, (c\,\varepsilon\,b \rightarrow c\,\varepsilon\,a) \wedge \forall e, f\, (e\,\varepsilon\,b \wedge f\,\varepsilon\,b \rightarrow e = f)] \quad (4.13)$$

The equivalence proof is not too complicated, so as an example I will only give it in one direction (4.11 $\Rightarrow$ 4.13). To complete the proof we only need to show that the right-hand side of (4.12) entails the right-hand side of (4.13):

---

[20] Remember, '$a = b$' stands here for '$a\,\varepsilon\,b \wedge b\,\varepsilon\,a$'.

1	$\exists c\,[\forall d\,(d\,\varepsilon\,b \equiv d = c) \land c\,\varepsilon\,a]$	Assumption
2	$\forall d\,(d\,\varepsilon\,b \equiv d = c) \land c\,\varepsilon\,a$	Assumption
3	$\forall d\,(d\,\varepsilon\,b \equiv d = c)$	CL: 2
4	$c\,\varepsilon\,a$	CL: 2
5	$c\,\varepsilon\,b \equiv c = c$	CL$\forall$: 3
6	$c\,\varepsilon\,c$	CL$\forall$: 4, (4.9)
7	$c = c$	Def: 6
8	$c\,\varepsilon\,b$	CL: 5, 7
9	$\exists c\,c\,\varepsilon\,b$	$\exists$Intro: 8
10	$e\,\varepsilon\,b$	Assumption
11	$e\,\varepsilon\,b \equiv e = c$	CL$\forall$: 3
12	$e = c$	CL: 10, 11
13	$e\,\varepsilon\,a$	4, 12
14	$\forall c\,[c\,\varepsilon\,b \to c\,\varepsilon\,a]$	10$\Rightarrow$ 13
15	$e\,\varepsilon\,b \land f\,\varepsilon\,b$	Assumption
16	$e\,\varepsilon\,b \equiv e = c$	CL$\forall$: 3
17	$f\,\varepsilon\,b \equiv f = c$	CL$\forall$: 3
18	$e = c \land f = c$	CL: 15, 16, 17
19	$e = f$	CL: 18
20	$\forall e, f\,(e\,\varepsilon\,b \land f\,\varepsilon\,b \to e = f)$	15$\Rightarrow$19
21	$\exists c\,c\,\varepsilon\,b \land \forall c\,[c\,\varepsilon\,b \to c\,\varepsilon\,a] \land \forall e, f\,(e\,\varepsilon\,b \land f\,\varepsilon\,b \to e = f)$	CL: 9, 14, 20
22	$\exists c\,c\,\varepsilon\,b \land \forall c\,[c\,\varepsilon\,b \to c\,\varepsilon\,a] \land \forall e, f\,(e\,\varepsilon\,b \land f\,\varepsilon\,b \to e = f)$	$\exists$Elim: 1, 2$\Rightarrow$21

4.7 Time Slices and Ontology

Leśniewski did not discuss three- or four–dimensionalism explicitly.[21] Nor did he consider time as a dimension while discussing mereology (which is understandable, given that he meant mereology to underlie mathematics rather than some more elaborate metaphysical theories). However, at least at one spot one can see that he was

---

[21] Three-dimensionalism is the view that what we naturally and primarily take to be objects are things that have three spatial dimensions. Four-dimensionalism on the other hand claims that it is the whole spatio-temporal 'worms' that we intuitively consider objects. For instance, on the three-dimensionalist view Rafal Urbaniak at this instant and Rafal Urbaniak from five minutes ago are not literally exactly the same object (although, there is a certain relation between them that resembles identity). On the four-dimensionalist view, Rafal Urbaniak at this instant is only part of the whole Rafal Urbaniak which is the whole consisting of Rafal Urbaniak at all moments at which he is alive.

conscious of some problems related to identity, time-slices etc. The discussion occurs in the context of Ontology, and the following sophism is discussed. Let us start with the following premises:

$$\text{Warsaw } \varepsilon \text{ older than the Saski Garden.} \tag{4.14}$$

$$\text{Warsaw of 1830 } \varepsilon \text{ smaller than Warsaw of 1930.} \tag{4.15}$$

$$\text{Warsaw of 1830 } \varepsilon \text{ Warsaw.} \tag{4.16}$$

$$\text{Warsaw of 1930 } \varepsilon \text{ Warsaw.} \tag{4.17}$$

'Older than the Saski Garden', 'smaller than Warsaw of 1930' and 'Warsaw' are all taken to be names, in Leśniewski's wide sense of the word. (The first one names all and only those things which are older than the Saski Garden, the second one all and only those things which are smaller than Warsaw of 1930.) Now, note that the following two are theorems of Ontology:

$$\forall a, b, c \, (a \, \varepsilon \, b \wedge c \, \varepsilon \, a \to c \, \varepsilon \, b) \tag{4.18}$$

$$\forall a, b, c, d \, (a \, \varepsilon \, b \wedge c \, \varepsilon \, a \wedge d \, \varepsilon \, a \to c \, \varepsilon \, d) \tag{4.19}$$

(4.19) entails:

$$\text{Warsaw } \varepsilon \text{ older than the Saski Garden } \wedge \text{ Warsaw of 1930 } \varepsilon \text{ Warsaw } \wedge$$
$$\wedge \text{ Warsaw of 1830 } \varepsilon \text{ Warsaw } \to \text{ Warsaw of 1930 } \varepsilon \text{ Warsaw of 1830} \tag{4.20}$$

But the antecedent is provided by (4.14), (4.16), and (4.17), so we get:

$$\text{Warsaw of 1930 is Warsaw of 1830.} \tag{4.21}$$

now, by applying (4.18) to (4.15) and (4.17) we conclude:

$$\text{Warsaw of 1930 is smaller than Warsaw of 1930.} \tag{4.22}$$

Since the conclusion is quite obviously false while the premises seem true, we need to find the fallacy. For Leśniewski, it is an equivocation of the word 'Warsaw' in the premises. To make (4.14) true, 'Warsaw' has to be taken as a singular term which names one object (presumably, Warsaw from the beginning to the end of its existence). But then, to time-slice of Warsaw (Warsaw of 1830, Warsaw of 1930) can be truly called Warsaw, and so (4.16) and (4.17) are false in this reading.

On the other hand, to make either (4.16) or (4.17) true, 'Warsaw' has to be taken as a general name referring to (at least two, if not all) time-slices of Warsaw. But if this is the case, 'Warsaw' cannot be a subject of a true atomic sentence and (4.14) is false.[22]

4.8 Leśniewski's Arithmetic

Using the language of Ontology, Leśniewski developed his own axiomatization of arithmetic (Srzednicki and Stachniak 1988, pp. 129–152). More or less, it is a version of Peano Arithmetic. The axioms are:

$$1 \, \varepsilon \, nat$$
$$\forall a \, [a \, \varepsilon \, nat \rightarrow sq(a) \, \varepsilon \, nat]$$
$$\forall a \, [a \, \varepsilon \, nat \rightarrow \neg sq(a) = 1]$$
$$\forall a, b \, [a \, \varepsilon \, nat \wedge b \, \varepsilon \, nat \wedge sq(a) = sq(b) \rightarrow a = b]$$
$$\forall a, b \, [1 \, \varepsilon \, b \wedge \forall c \, (c \, \varepsilon \, nat \wedge c \, \varepsilon \, b \rightarrow sq(c) \, \varepsilon \, b) \wedge a \, \varepsilon \, nat \rightarrow a \, \varepsilon \, b]$$
$$\forall a \, [a \, \varepsilon \, nat \rightarrow a + 1 = sq(a)]$$
$$\forall a, b \, [a \, \varepsilon \, nat \wedge b \, \varepsilon \, nat \rightarrow a \times (b + 1) = (a + b) + 1]$$
$$\forall a \, (a \, \varepsilon \, nat \rightarrow a \times 1 = a)$$
$$\forall a, b \, [a \, \varepsilon \, nat \wedge b \, \varepsilon \, nat \rightarrow a \times (b + 1) = (a \times b) + a]$$
$$\forall a, b \, [a \, \varepsilon \, nat \wedge b \, \varepsilon \, nat \wedge a > b \rightarrow \exists c \, (c \, \varepsilon \, nat \wedge a = b + c)]$$
$$\forall a, b \, [a \, \varepsilon \, nat \wedge b \, \varepsilon \, nat \rightarrow a + b > a]$$
$$\forall a, b \, (a \, \varepsilon \, sq(b) \rightarrow b \, \varepsilon \, nat)$$

All but three last axioms are fairly standard axioms of (second order) Peano arithmetic, including the second order induction axiom. The third and second last axioms are added to handle ordering. Only the last axiom is somewhat unusual, because the lecture notes suggest that "it is not a thesis of Peano arithmetic" and that it is essential for proving $\forall a, b \, (a \, \varepsilon \, sq(b) \rightarrow a \, \varepsilon \, nat)$.[23] The lecture notes claim

---

[22] Something that Leśniewski didn't say (but which corresponds to the four-dimensionalist view) is that there is a reading which makes all premises true but which makes the application of (4.18) and (4.19) illegitimate, because the logical form of (4.16) and (4.17) is not the same as that of $a \, \varepsilon \, b$. In this reading 'Warsaw' refers to a unique, four-dimensional object (Warsaw from the beginning to the end of its existence), 'Warsaw of 1830' refers to a yearly time-slice of Warsaw and 'Warsaw of 1930' refers to another yearly time-slice of Warsaw. (4.16) instead of saying that among object(s) named by 'Warsaw' there is the object which is named by 'Warsaw of 1830' says rather that Warsaw of 1830 is part of Warsaw (and similarly for (4.17)). But in this reading it does not follow that Warsaw of 1930 is Warsaw of 1830, even if 'Warsaw' names exactly one object and (4.14) is true.

[23] "The axiom ... restricts the scope of the term 'sq'. Without this restriction the axiom system of Peano arithmetic is not sufficient." (Srzednicki and Stachniak 1988, p. 141)

independence, but do not provide any independence proof. Leśniewski then proceeds to meticulously derive quite a few moderately exciting consequences, such as the commutativity, associativity and distributivity of arithmetical operations.[24]

4.9 Further Readings

The first thing the reader may want to take a look at is the only existing collection of papers related directly to Leśniewski's Ontology and Mereology (Srzednicki and Rickey 1984). It contains English translations of Kruszewski (1925) and Sobociński (1949, 1954). Lejewski (1954, 1958, 1969), Słupecki (1955), Canty (1969), Iwanuś (1973), and Clay (1966, 1970, 1974a) are also included.

The simplest subsystem of Ontology we get when we allow only name variables and do not admit any quantification. A general account and metatheory of calculi of this kind can be found in Pietruszczak (1991). Pietruszczak constructs various calculi of this kind and provides a general method for proving completeness of those with respect to set-theoretic semantics. Since the book was published only in Polish his work is relatively unknown. Quite independently, a subsystem of Ontology of this sort (quantifier–free, name variables only), called L_1 was studied by Ishimoto (1977); Ishimoto and Kobayashi (1982); Ishimoto (1997).

More general systems are generated by introducing quantifiers binding name variables only. Some authors call the subsystem of Ontology which contains only this sort of quantifiers 'elementary Ontology'. Some other authors use 'elementary Ontology' to refer to a rather stronger system which admits also second-order variables and quantifiers binding them (that is, we also allow variables representing functors that take name arguments only), but no variables (or constants) of other categories.Elementary Ontology in the latter sense is discussed especially by Iwanuś (1973)—who also shows it to be equivalent to the theory of atomic Boolean algebras and the classical theory of set algebras. Iwanuś also provides an interesting decidability proof for Elementary Ontology.

Elementary Ontology is also discussed in the first part of one of the best surveys on Ontology (Słupecki 1955). Słupecki, however, discusses also some aspects of non-elementary Ontology, which is obtained mainly by adding the rule of extensionality. He proves that definitions in the system satisfy the condition of translatability, i.e., for every ϕ which contains a defined term there is a provably equivalent expression ϕ' which does not contain this term. Interestingly, definitions are creative in the system. They can be made non-creative by introducing two axiom schemata. Słupecki also sketches the proof of the theorem that Ontology is consistent relative to Protothetic.

A specific type of functional completeness of Elementary Ontology with respect to ε has been proven by Urbaniak (2006a).

---

[24] Leśniewski also attempted to eliminate recursive definition by means of explicit definitions in the language of Ontology. Alas, the notes on this topic are rather scant and the issue requires further investigation beyond the scope of this book.

Lejewski (1958) is intended as a survey of the full–blown Ontology. It contains more elaborate intuitive explanations of the intended meaning of epsilon and some other connectives definable in Ontology. Lejewski, who is interested in replacing epsilon with another, more intuitive for native English speakers, connective proves that Ontology can be constructed with inclusion as the sole primitive term (Lejewski, 1977).[25]

Various equivalent axiomatizations of Ontology are discussed in Sobociński (1934). Miéville (2004) is a recent French text intended as a historical introduction to Ontology.

Canty (1969) explains the relation between the epsilon, the distributive predication, and higher-order epsilon connectives. The discussion is quite informal.

Urbaniak (2006b) proves that in set-theoretic semantics the classical Ontology has at least $\aleph_0$ non-standard models with a non-standard ε. Urbaniak (2009) strengthens the result by proving that no set of second-order formulas of the language of Ontology can unambiguously fix the set-theoretic interpretation of ε.

Kubiński (1969) gives a convincing sketch of a proof that Ontology is absolutely non-categorical (i.e. that no set of formulas of Ontology defines a class of models up to isomorphism).

Leśniewski's systems raise also certain philosophical concerns. One of the crucial questions (which I discuss in Chap. 8) is the meaning of quantifiers and variables. Leśniewski claimed that his systems are nominalistic but he used something that nowadays can be easily construed as higher-order quantification. How to interpret the role of variables and quantifiers in Leśniewski's systems to avoid this clash?

Another interesting aspect is how Ontology can be applied in philosophy. Does formalizing arguments using Ontology provide us with a new insight into their quality or meaning?

Munitz (1974) is an interesting book that covers some differences in expressing existence in various logical systems, Ontology included. Existential assumptions and commitments as expressed in Ontology were, it seems, first discussed in Lejewski (1954). An interesting commentary to Lejewski is Kearns (1969).

Prior (1965) argues that variables in Ontology represent class names rather than singular or empty or common terms. However, he does not explain what the difference is and why it is so important. He does not explain what he means by a class name.[26] Sagal (1973) criticizes Prior.[27]

Since the quantifiers in Leśniewski's systems do not (or at least, were not intended to) express ontological commitment, a perfectly legitimate question arises as to what those quantifiers were supposed to do and how we are to interpret variables that occur in the system. For instance, there is some similarity between Leśniewski's account

---

[25] This inclusion functor in the classical formulation of Ontology is defined by (4.6) on p. 87.

[26] Also, the main argument that he presents is not very compelling. It seems to be this: regular names cannot be logically complex (he does not explain this notion either), class names can, names in Ontology can be complex, therefore names in Ontology are class names.

[27] He does so mostly on the grounds that this reading disagrees with Leśniewski's dislike of abstract objects.

and the theory of plural quantification developed by Boolos (1998a) (informally discussed by Simons (1997)).[28]

The question of ontological commitment of Ontology is also raised by Simons (1995). Simons suggests a different criterion of ontological commitment (which is, unfortunately quite vague):

> I shall say that a set of propositions ...ontologically commits to those objects (of whatever ontological category) that, in some sense of 'exist' have to exist if all the propositions in the set are true.

and argues that Ontology has no ontological commitments. His claim that Ontology is valid if the domain of individuals is empty is certainly true (he gives a proof). However, this does not seem to answer exactly the question whether it is free of any commitments. In his proof, the truth-conditions of higher-order statements tacitly do much of the heavy lifting, and the question whether their use does not involve any commitment has not been successfully answered. Simons attempts to provide some hints in Simons (1985b), where he suggests his 'combinatorial semantics' in which name variables (are supposed to) range over 'ways of meaning', but it is not a precise account and this approach has not yet been fully developed and assessed. He does pose an interesting problem: how should one construct a formal semantics for Leśniewski's logic which performs the task a formal semantics normally performs, but does not assume explicitly or implicitly the existence of abstract objects?

Küng and Canty (1970) argue that the quantification in Ontology is not objectual (referential).[29] The authors claim that the interpretation should be substitutional (although they suggest that they can be interpreted as referential over a realm of sets associated with a domain of objects). Kielkopf (1977) criticizes the view in its whole generality, but he introduces some other subtle interpretations of quantifiers which are neither straightforward substitutional nor referential. Küng (1977) applies

---

[28] There are of course some differences. For Leśniewski, quantification does not express existence (i.e. $\exists a\, \phi$, philosophically is not to be meant as stating that there exists a such that ϕ). For Boolos individual quantification commits one to individuals and the higher-order quantification in the standard interpretation commits one to certain kinds of abstract entities, but monadic higher-order quantification interpreted as relating to plurals commits us to only whatever individuals we need to render the relevant sentence true. Also, Leśniewski assumes that the category of names contains singular names, empty names and plurals indiscernibly. Boolos treats plurals as another group in its own right. He has a syntactic distinction between two kinds of nominal variables. I discuss the relation in more detail in Chap. 8.

[29] The quantification in a given language is called referential (or objectual) with respect to a given semantics of this language if the truth in a model of quantified statements is dependent upon there being (or not) some objects in the domain of this model satisfying (or not) the formula resulting from the initial formula by deleting the quantifier. Or, in other words, in the objectual interpretation variables in a model are assigned objects in the domain of this model. The objectual reading is often opposed to the substitutional reading, where variables are rather associated with a substitution class—the class of expressions (which do not belong to the domain of the model), and the satisfaction of a formula in a model is defined via some results of substitution being true (or not). Clearly, Ontology as an axiomatic system can be given both an objectual and a substitutional interpretation. Presumably, the question of whether quantification in Ontology is objectual or substitutional boils down to something like: 'which interpretation is in accordance with Leśniewski's views?'.

so-called prologue-functors (developed by Küng (1974)) like 'whatever extension
the inscriptions equiform with the following item are taken to have ...the following
is asserted ...' to account for quantification in Leśniewski's systems (whether this
reading is nominalistic depends on how the notion of extension is understood,and
providing a nominalistic explanation of what an extension is is not an easy matter).

A very interesting approach to semantics of Ontology has been taken by Rickey
(1985). Quantified name calculi are interpreted in first-order two-sorted models. One
sort is the sort of individuals, another sort is the sort of names. A binary relation on
the Cartesian product of those two sorts is added, which intuitively corresponds to
the relation of naming.

Sobociński (1949) deals with Leśniewski's solution to Russell's paradox
(Sobociński 1949 was published in a few parts, but is reprinted in whole in
Srzednicki and Rickey 1984). There are some reasons to think that this solution
is not satisfactory, as I argue in detail in Chap. 7 (see also Urbaniak 2008).

Lejewski (1985) is an interesting attempt to use the expressivity of the language
of Ontology to capture main intuitions pertaining to sets. I discuss this attempt and
indicate its weaknesses in section also in Chap. 7.

Definite descriptions as formulated in Leśniewski's Ontology have also been studied. Hiż (1977) proved that the definition of descriptions in ∗14.01 of *Principia
Mathematica* when formulated in the language of Ontology, is in Ontology provably
equivalent to the axiom of Ontology. I described this result in Sect. 4.6. Another
discussion of Russell's theory of descriptions from a Leśniewskian point of view is
Lejewski (1960), to which Russell responded in Russell (1960).

Lambert and Scharle (1967) argue that the system of free logic FL^t is trans-
latably equivalent to a first-order modification of elementary Ontology called $L4'$.
The authors suggest that their translation "Provides for the first time ...a way of
interpreting at least the first order fragment of one version of Leśniewski's system
called Ontology in more general parlance". Simons (1981) correctly criticizes this
claim, where he points out that $L4'$ (1) is not a subsystem of Ontology, because it
has an additional axiom requiring that no term be general, and (2) even if it were a
subsystem, it would be a narrow part of Ontology, and so it is false that free logics
provide *a more general parlance* to speak of Ontology. Interesting modifications of
Ontology that allow to embed some free logics are developed by Simons (1985a).

Grzegorczyk (1955) compares Ontology to Boolean algebra and suggests that
Ontology formally resembles the theory of complete atomic Boolean algebras with
constants and functions of an arbitrary high type. Iwanuś (1973), p. 202 elaborates
on this issue, emphasizing some differences: Boolean algebras are not provided with
a rule of definitions, so the statement should be conditional: if we extend the theory of Boolean algebras with a good account of definitions (which is pretty much
a translation of Leśniewski's rules), then there is a correspondence. Also, the proof
of the equivalence between elementary Ontology (in the stronger sense) and theory
of Boolean algebras states something about a proper subsystem of Ontology only.
Interestingly, it was Leśniewski's research that inspired Tarski's work on the foundations of Boolean algebras (1935), which played an essential role in the development
of Boolean algebras.

Stachniak (1981) develops a strengthening of Ontology, L_{DF} and a semantics for this calculus, where models are taken to be atomic Booleanalgebras enriched by a specific set of functions and relations. The strengthening consists in including an axiom schema which, briefly speaking, resembles the axiom of comprehension. Stachniak uses a fairly standard method of constructing models from constants to prove the completeness and compactness theorems. He also discusses the ultraproduct construction of models for Ontology. Definitions in L_{DF} are not creative.

Lebiedewa (1969a; 1969b) extends Ontology by introducing modalities.

There is also an interesting modification of Ontology which is meant to deal with vague terms (Kubiński, 1958). Kubiński (1959) developed also a modification of Ontology that is intended to deal with apparent contradictions of the form 'a is b and a is not b'.Unfortunately, both papers are available in Polish only. Kubiński (1960) extends Ontology with means to express something like 'a is more like b than like c'.

An ontological version of the axiom of choice is independent of Ontology. Kowalski (1977) studies the extension of Ontology obtained by adding this axiom, and Davis (1975) discusses various formulations of this axioms. Perhaps the most elegant of these formulations is: '$\exists f\, \forall a,\, g\,(g(a) \to g(f(a)))$'.

An interesting extension of Ontology is constructed when we add an axiom of infinity. When this is done, a theory that bi-interprets Peano arithmetic is derivable (i.e. is a proper sub theory of the theory thus obtained). Gödel's theorem applies to this extension (Canty 1967). In a sense, those strengthenings can constitute foundational systems that Leśniewski was after. There are some problems, however,which mostly have to do with Leśniewski's philosophical views. Leśniewski claimed to be a nominalist. Unless a plausible nominalist story about quantification in Ontology is told, it is fairly unclear that Ontology itself is a nominalistically acceptable tool. Leśniewski also opposed the idea of modifying a system "just to make it work" (that was his main objection against developing weaker set theories just to avoid the Russell's paradox) and he believed that the axioms of his systems correspond to certain essential features of reality (he even called them true). That being the case, a question arises as to whether postulating the actual existence of infinitely many objects is acceptable in this setting.

References

Boolos, G. (1998a). Nominalist platonism. In R. Jeffrey (Ed.), *Logic, logic, and logic* (pp. 73–87). Harvard University Press, Cambridge, Massachusetts, London.

Canty, J. T. (1967). Leśniewski's Ontology and Gödel's Incompleteness Theorem. PhD thesis, University of Notre Dame.

Canty, J. T. (1969). Ontology: Leśniewski's logical language. *International Journal of Language and Philosophy, 5*, 455–469.

Clay, R. (1966). On the definition of mereological class. *Notre Dame Journal of Formal Logic, 7*(4), 359–360.

Clay, R. (1970). The dependence of a mereological axiom. *Notre Dame Journal of Formal Logic, 11*(4), 471–472.

Clay, R. (1974a). Relation of Leśniewski's mereology to boolean algebra. *The Journal of Symbolic Logic*, *39*(4), 638–648.

Davis, C. (1975). An investigation concerning the Hilbert-Sierpiński logical form of the axiom of choice. *Notre Dame Journal of Formal Logic*, *16*, 145–184.

Gryganiec, M. (2000). Leśniewski przeciw powszechnikom [Leśniewski against universals]. *Filozofia Nauki*, *3–4*, 109–125.

Grzegorczyk, A. (1955). The systems of Leśniewski in relation to contemporary logical research. *Studia Logica*, *3*, 77–95.

Hiż, H. (1977). Descriptions in Russell's theory and Ontology. *Studia Logica*, *36*(4), 271–283.

Ishimoto, A. (1977). A propositional fragment of Leśniewski's ontology. *Studia Logica*, *36*, 285–299.

Ishimoto, A. (1997). Logicism revisited in the propositional fragment of Leśniewski's ontology. In E. Agazzi, & G. Darvas (Eds.), *Philosophy of mathematics today* (pp. 219–232). Volume 22 of Episteme, Kluwer, Netherlands.

Ishimoto, A., & Kobayashi, M. (1982). A propositional fragment of Leśniewski's ontology and its formulation by the tableau method. *Studia Logica*, *41*, 181–195.

Iwanuś, B. (1973). On Leśniewski's elementary ontology. *Studia Logica*, *31*, 7–72.

Kearns, J. (1969). Two views of variables. *Notre Dame Journal of Formal Logic*, *10*, 163–180.

Kielkopf, C. (1977). Quantifiers in Ontology. *Studia Logica*, *36*, 301–307.

Kowalski, J. (1977). Leśniewski's Ontology extended with the axiom of choice. *Notre Dame Journal of Formal Logic*, *18*(1), 1–78.

Kruszewski, Z. (1925). Ontologia bez aksjomatów [Ontology without axioms]. *Przegld Filozoficzny*, *28*, 136 [abstract].

Kubiński, T. (1958). Nazwy nieostre [Vague names]. *Studia Logica*, *7*, 115–176.

Kubiński, T. (1959). Systemy pozornie sprzeczne [Apparently inconsistent systems]. *Zeszyty Naukowe Uniwersytetu Wrocławskiego*, 53–61.

Kubiński, T. (1960). An attempt to bring logic nearer to colloquial language. *Studia Logica*, *10*, 61–75.

Kubiński, T. (1969). Pewna teoriomnogościowa własność Ontologii [A certain set-theoretic property of Ontology]. *Ruch Filozoficzny*, *27*(4).

Küng, G., & Canty, J. T. (1970). Substitutional quantification and Leśniewskian quantifiers. *Theoria*, *36*, 165–182.

Küng, G. (1974). Prologue-functors. *Journal of Philosophical Logic*, *3*, 241–254.

Küng, G. (1977). The meaning of the quantifiers in the logic of Leśniewski. *Studia Logica*, *36*, 309–322.

Lambert, K., & Scharle, T. (1967). A translation theorem for two systems of free logic. *Logique et Analyse*, *10*, 328–341.

Lebiediewa, S. (1969a). The systems of modal calculus of names I. *Studia Logica*, *24*, 83–104.

Lebiediewa, S. (1969b). Systemy Modalnego Rachunku Nazw Nadbudowane nad Rachunkiem Nazw Stanisława Leśniewskiego. PhD thesis, Uniwersytet Wrocławski.

Leśniewski, S. (1913). Krytyka logicznej zasady wyłączonego środku. *Przegld Filozoficzny* [Philosophical Review], *16*, 315–352. [The critique of the logical principle of excluded middle, (Leśniewski 1991, 47–85)].

Leśniewski, S. (1927). O Podstawach Matematyki, Wstęp. Rozdział I: O pewnych kwestjach, dotyczczych sensu tez 'logistycznych'. Rozdział II: O 'antynomji' p. Russella, dotyczcej 'klasy klas, nie będcych własnemi elementami'. Rozdział III: O różnych sposobach rozumienia wyrazów 'klasa' i 'zbiór'. *Przegld Filozoficzny*, *30*, 164–206. [On the foundations of mathematics. Introduction. Ch. I. On some questions regarding the sense of the 'logistic' theses. Ch. II. On Russel's 'antinomy' concerning 'the class of classes which are not elements of themselves'. Ch. III. On various ways of understanding the expression 'class' and 'collection' (Leśniewski 1991, 174–226)].

Leśniewski, S. (1931). O podstawach matematyki, Rozdział X: Aksjomatyka 'ogólnej teorji mnogości pochodzca z r. 1921. Rozdział XI: O zdaniach 'jednostkowych' typu '*Aεb*. *Przegld*

Filozoficzny, 34, 142–170. [On the foundations of mathematics. Ch. X. The axiomatization of the 'general theory of sets' from the year 1921. Ch. XI. On 'singular' propositions of the tyle 'Ab, (Leśniewski 1991, 350–382)].

Lejewski, C. (1954). Logic and existence. *The British Journal for the Philosophy of Science, 5*, 104–119.

Lejewski, C. (1958). On Leśniewski's Ontology. *Ratio, 1*(2), 150–176.

Lejewski, C. (1960). A re-examination of the Russellian theory of descriptions. *Philosophy, 35*, 14–29.

Lejewski, C. (1969). Consistency of Leśniewski's mereology. *The Journal of Symbolic Logic, 34*(3), 321–328.

Lejewski, C. (1977). Systems of Leśniewski's Ontology with the functor of weak inclusion as the only primitive term. *Studia Logica, 36*, 323–349.

Lejewski, C. (1985). Accommodating the informal notion of class within the framework of Leśniewski's Ontology. *Dialectica, 39*, 217–241.

Miéville, D. (2004). *Introduction à l'oeuvre de S. Lesniewski. Fascicule II: L'ontologie.* Neuchâtel: Centre de Recherches Sémiologique.

Munitz, M. (1974). *Existence and Logic.* New York: New York University Press.

Pietruszczak, A. (1991). Bezkwantyfikatorowy Rachunek Nazw. Systemy i ich Metateoria [Quantifier-free Name Calculus. Systems and their Metatheory]. Adam Marszałek, Toruń.

Prior, A. (1965). Existence in Leśniewski and Russell. In J. Crossley (Ed.), *Formal systems and recursive functions* (pp. 149–155). Amsterdam: North-Holland.

Rickey, F. (1985). Interpretations of Leśniewski's Ontology. *Dialectica, 39*(3), 181–192.

Russell, B. (1960). Response to Lejewski's criticism. *Philosophy, 35*, 146.

Sagal, P. (1973). On how best to make sense of Leśnicwski's Ontology. *Notre Dame Journal of Formal Logic, 14*(2), 259–262.

Simons, P. (1981). A note on Leśniewski and free logic. *Logique et Analyse, 24*, 415–420.

Simons, P. (1985a). Lesniewski's logic and its relation to classical and free logic. In G. Dorn, & P. Weingarten (Eds.), *Foundations of logic and linguistics. Problems and solutions* (pp. 369–400). *Plenum, New York.*

Simons, P. (1985b). *A semantics for Ontology. Dialectica, 39*(3), 193–215.

Simons, P. (1995). Lesniewski and ontological commitment. In D. Miéville, and D. Vernant (Eds.), Stanislaw Lesniewski Aujourd'hui, number 16 in Recherches Philosophie, Langages et Cognition (pp.103–119). Université de Grenoble.

Simons, P. (1997). Higher-order quantification and ontological commitment. *Dialectica, 51*, 255–271.

Słupecki, J. (1955). St. Leśniewski's calculus of names. *Studia Logica, 3*, 7–72.

Sobociński, B. (1934). O kolejnych uproszczeniach aksjomatyki Ontologii prof. St. Leśniewskiego [On subsequent simplifications of the axiomatics of prof. Leśniewski's Ontology]. *Fragmenty Filozoficzne,* 143–160.

Sobociński, B. (1949). L'analyse de l'antinomie Russellienne par Leśniewski. *Methodos, 1–2*(1, 2, 3; 6–7):94–107, 220–228, 308–316; 237–257. [translated as "Leśniewski's analysis of Russell's paradox" (Srzednicki and Rickey 1984,11–44)].

Sobociński, B. (1954). Studies in Leśniewski's Mereology. *Yearbook for 1954–55 of the Polish Society of Arts and Sciences Abroad, 5*, 34–48.

Srzednicki, J., & Stachniak, Z. (Eds.). (1988). *S.Leśniewski's Lecture Notes in Logic.* Dordrecht: Kluwer Academic Publishers.

Srzednicki, J., & Rickey, F. (Eds.). (1984). *Leśniewski's Systems.* Ossolineum, The Hague, Wrocław: Ontology and Mereology. Martinus Nijhow Publishers.

Stachniak, Z. (1981). *Introduction to Model Theory for Lesniewski's Ontology.* Wrocław: Wydawnictwo Uniwersytetu Wrocławskiego.

Tarski, A. (1935). Zur Gundlegung der Boole'schen llgebra. *Fundamenta Mathematicae, 24*, 177–198.

Urbaniak, R. (2006a). On Ontological functors of Leśniewski's elementary Ontology. *Reports on Mathematical Logic*, *40*, 15–43.

Urbaniak, R. (2006b). Some non-standard interpretations of the axiomatic basis of Lesniewski's Ontology. *The Australasian Journal of Logic*, *4*, 13–46.

Urbaniak, R. (2008). Leśniewski and Russell's paradox: Some problems. *History and Philosophy of Logic*, *29*(2), 115–146.

Urbaniak, R. (2009). A note on identity and higher-order quantification. *The Australasian Journal of Logic*, *7*, 48–55.

Waragai, T. (1981). Leśniewski's refutation of general object on the basis of ontology. [*Journal of Gakugei, 30*

Chapter 5
Leśniewski's Mereology

Abstract First, I focus on Leśniewski's philosophical motivations for Mereology: he constructed the system to provide a nominalistically acceptable alternative to set theory as a foundational system. Next, I compare axiomatizations from years 1916, 1918, 1920 and 1921. Then, I introduce a selection of mereological theorems. Some seem to support the claim that Mereology can be a sensible replacement for set theory and some indicate the extent Mereology to which it is different from the standard set theory. I also critically assess the potential role of Mereology in the foundations of mathematics. My assessment is rather negative. (It will be clear in chaps. 7 and 8 that it is Ontology and not Mereology that should play the foundational role.) The last section contains a survey of secondary literature related to Mereology.

5.1 Why Mereology?

Mereology[1] was the first deductive system developed (around 1916) by Leśniewski.[2] It arose from his considerations regarding Russell's antinomy.[3] He wanted to provide an alternative system of foundations of mathematics which avoids antinomies, but

---

[1] The name originates from the ancient Greek and loosely speaking means 'a science about parts'.

[2] So, historically speaking, it was the first theory he constructed, however imperfect formally it was at that time. In his endeavor to provide it with a secure logical background theory he later developed Ontology and Protothetic. Goodman in *The structure of appearance* (Goodman 1966) credits Leśniewski with the publication of "*a* [emph. mine] calculus of individuals". The one Goodman presents in the book is a variant formulated by Leonard in *The calculus of individuals and its uses* (Goodman and Leonard 1940). The mereological axioms are equivalent to Leśniewski's, but the underlying logic in Goodman and Leonard (1940) is weaker.

[3] "I occupied myself zealously with the 'antinomies'. From the time when in the year 1911 I began an acquaintance with them by meeting with the 'antinomies' of Russell related to the 'class of classes not elements of themselves', and problems concerning the antinomies were the most demanding subject of my deliberations for over eleven years." (Leśniewski 1927, 199)

R. Urbaniak, *Leśniewski's Systems of Logic and Foundations of Mathematics*,
Trends in Logic 37, DOI: 10.1007/978-3-319-00482-2_5,
© Springer International Publishing Switzerland 2014

unlike standard set theory does not pay the prize of giving up on intuitively compelling assumptions. In the preface to the first presentation of his Mereology he writes:

> The present work is the first link in an extended series of works, which I intend to publish in the near or distant future, desiring to contribute as much as possible to the justification of modern mathematics ...The arrangement of definitions and truths, which I established in the present work dedicated to the most general problems of the theory of sets, has for me, in comparison to other previously known arrangements of definitions and truths (Zermelo, Russell, etc.) this advantage that it eliminates the 'antinomies' of the general theory of sets without narrowing the original domain of Cantor's term 'set' ...and on the other hand, it does not lead to assertions which are in such startling conflict with intuitions of the 'commonalty'...(Leśniewski 1916, 129–130)

The basic idea behind Mereology is that a collection of material objects is nothing more than a concrete whole consisting of those objects taken together. This means, for example, that the mereological class of stones in a given heap is nothing more than the heap itself. In the early stage of development of set-theory the informal explanations of what a set is indeed might have suggested this reading. For example, Cantor says:

> Any set of distinct things can be regarded as a single thing in which those objects are constituents or constitutive elements.[4]

For Cantor, examples of objects standing in the relation of constitution are: (a) particular sounds and the whole musical composition, (b) a picture and its suitably matched parts. Considering the way the development of Cantor's set-theory went it seems that Cantor did not understand sets mereologically. The point is, however, that in the early years of set theory, informal explanations were not precise enough to exclude this interpretation.

As we have already seen (in Sect. 2.8.4), Leśniewski relied on mereological intuitions about classes already when he dealt with Russell's paradox in (1914a). At that time, however, he did not use any axiomatization of his theory of parthood. Only in 1916 he took the first stab at constructing such a system. It was not technically perfect. There were axioms, definitions, theorems and proofs, but he did not go far with formalization. He did use something like name variables, but everything else was stated using natural language. Also, the logic underlying his proofs was not explicitly formulated. Strictly speaking, while publishing on Mereology, Leśniewski never went much further in the level of formalization.[5] Only in later papers did he fill

---

[4] "Jede Menge wohlunterschiedener Dinge kann als ein einheithliches Ding für sich angesehen werden, in welchem jene Dinge Bestandteile oder constitutive Elemente sind." (my translation of Cantor 1877, 83)

[5] In his last paper on Mereology he does suggest abbreviating some name-forming functors, but the proofs and theorems are still stated quite informally (Leśniewski 1931a). Still, it deserves the name of a formalization because he introduces axioms and definitions, which despite containing certain natural language expressions, like 'is a part of' or 'is' are formulated in quite a regimented language. He also proves that certain other expressions which he calls theorems follow from these axioms and definitions. Even though he had not formulated clearly his rules of inference by that time, his proofs were quite regimented and detailed. In a way, they look very much like an applied

in the gap by introducing two logical systems (Protothetic and Ontology) in which his prior informal arguments can be formalized.

Leśniewski adhered to the view which identifies classes with what we nowadays would call 'mereological fusions' and he intended to build a few claims into his system: he rejected the existence of the empty set, did not distinguish between an object and its singleton, wanted to avoid reference to abstract objects and extensions, and disagreed with Russell and Whitehead's no class theory. Let's take a look at a closer look at these points.

First, he considered the existence of the empty set dubious. The empty set seemed for him to be something invented or introduced just for convenience (mainly to make the set product operation uniform and executable for any sets).[6] Also, Leśniewski took antorian definition seriously. If any set is constituted by its elements, then whenever there is nothing that constitutes it, the set itself does not exist.

> Being of the opinion that, if an object is the class of some a …then it actually consists of a, I always rejected …the existence of theoretical monstrosities like the class of square circles, understanding only too well that nothing can consist of something which does not even exist. (Leśniewski 1927, 214)

Second, he did not recognize the distinction between an object and its singleton. Here it is important to distinguish two readings of the claim that an object is identical with its singleton. In the first reading, accepted by Leśniewski, if a name m names exactly one object x, then the expression 'the class of m' also names exactly one object, namely x itself. However, Leśniewski did not accept the view that for any object x, x is a class k such that any object y which is an element of k is identical

---

(Footnote 5 continued)
system of natural deduction (e.g. to prove a conditional he assumes the antecedent and proves the consequent; to prove a conditional within the scope of a universal quantifier he freezes the relevant variable(s), assumes the antecedent and proves the consequent).

[6] For example, Fraenkel says: "On purely formal grounds, namely to be able to express certain facts more easily and more conveniently, we introduce at this point a further improper set, the so-called null-set. This is defined as containing no element whatever; it is therefore in virtue of the definition from page 4 not a set at all, but it will on the ground of a special appointment and by an extension of that definition, will be counted as one (indeed as a finite one) and will be designated by 0." (my translation of Fraenkel 1928, 21) Also in a more modern text Fraenkel, Bar-Hillel and Levy suggest that the reason for postulating the empty set is practical: "The practical reasons which call for the existence of an individual are as follows. When we define the intersection of two sets r and s to be the set t which consists of those elements which belong to both r and s, we want the intersection to be defined even in the case where r and s have no members in common. In this case the intersection t has to be a memberless element, i.e. an individual. There are also many other examples where the existence of a memberless element makes things simpler. The same practical reasons which call for the existence of such an element also call for using always the same element for the intersection of any two sets r and s with no members in common, and for referring to this element as a set. Therefore we shall call this element the null-set and our sets are, from now on, the elements which have members as well as the null set. Let us, however, stress at this point that whereas the existence of at least one individual is required for serious philosophical reasons, referring to one of the individuals as the null-set is done only for reasons of convenience and simplicity, and can be regarded as a mere notational convention."(Fraenkel et al. 1973, 24)

with the object x. In other words, he accepted the first and rejected the second of these claims:

$$m\varepsilon m \rightarrow Kl(m) = m$$

$$m\varepsilon m \rightarrow \forall b \, (b \text{ is an element of } Kl(m) \rightarrow b\varepsilon m)$$

The first claim is in accordance with the mereological account of classes. The second (insofar as we accept Leśniewski's identification of being an element with being a part[7]) goes against it. Let us name a specific chosen lizard 'Aldo'. Now, the singleton of Aldo is (for Leśniewski) Aldo himself. It is not the case, however, than any element of this singleton (that is, any part of Aldo) is Aldo.

This issue touches upon another interesting feature of mereological classes. They have cardinality only relative to countable noun phrases. The question 'how many objects does the heap of stones consist of?' is as difficult to answer as the question 'how many parts does this heap have?'. On the other hand, the question like 'how many stones does it consist of?' seems easier to answer.

Every mereological class can also be viewed as singleton, relative to the name of itself. So for example, the class of humans is the mereological whole composed of all people. But the class of the class of humans is a singleton, in the sense that it is the class of certain a (namely, 'the class of humans'), where a is a singular term. This also indicates that using the term 'singleton' in this loose, relative sense is not too useful.[8]

Third, Leśniewski neither believed in the existence of abstract objects nor did he think he understood in the slightest explanations that employed 'extensions of concepts', 'concepts' or 'courses of values of functions':

> I will not attempt here to submit Frege's conception which treats classes as 'extensions of concepts' to analytic assessment because, despite my best efforts in this direction, I am still unable to understand what the various authors are saying when they use the expression 'extension of a concept'. If the 'extension of concept a' is not to be the class of objects a consisting of objects a in accordance with my conception of classes, then, being unable to answer my own question as to what this 'extension of concept a' could be and when and where one could become acquainted with such an 'extension' and whether anything like it exists in the world, I am none the less inclined to surmise meekly that it is simply an object 'devised' by logicians for the annoyance of many generations. I am no more able to understand Frege's utterance …that the 'extension of a concept' "attaches to the concept itself and to it alone", than I am the most obscure enunciation of the exponents of 'romantic philosophy', which simply means that I do not understand the utterance at all. The declaration, according to which "With such functions whose value is always a truth-value, we can say 'extension of the concept' instead of 'course of values of the function' and it seems appropriate to call directly a concept, a function whose value is always a truth-value" does nothing to clarify for me the question of the 'extensions of concepts' because the expression 'course of values

---

[7] More technically, with being an ingredient. I will discuss the details of this identification later.

[8] One could try to define the notion differently and say that a singleton is a class which has a unique element. But in this sense, the fact that a class is the class of a's, where a is a singular term is not sufficient for this class to be a singleton (although it is a necessary condition). What is sufficient is that the object which generates the class has no parts, that is, is an atom.

of the function' is no more intelligible to me than the expression 'extension of the concept'. (Leśniewski 1927, 219–220)

Also, Leśniewski objected to the philosophical underpinnings of Whitehead's and Russell's no-class theory. Basically, the no-class theory in *Principia Mathematica* takes the symbols for classes to be incomplete symbols, just like definite descriptions, and provides a way of reducing expressions containing them to expressions which do not contain symbols for classes. A philosophical motivation for this approach to be found in *Principia Mathematica* is:

> The symbols for classes, like those for descriptions, are, in our system, incomplete symbols: their uses are defined, but they themselves are not assumed to mean anything at all. That is to say, the uses of such symbols are so defined that, when the definiens is substituted for the definiendum, there no longer remains any symbols which could be supposed to represent a class. Thus classes, so far as we introduce them, are merely symbolic or linguistic conveniences, not genuine objects as their members are if they are individuals.

> It is an old dispute whether formal logic should concern itself mainly with intensions or with extensions. In general, logicians whose training was mainly philosophical have decided for intensions, while those whose training was mainly mathematical have decided for extensions. The facts seem to be that, while mathematical logic requires extensions, philosophical logic refuses to supply anything except intensions. Our theory of classes recognizes and reconciles these two apparently opposite facts, by showing that an extension (which is the same as a class) is an incomplete symbol, whose use always acquires its meaning through a reference to intension.

> In the case of descriptions, it was possible to prove that they are incomplete symbols. In the case of classes, we do not know of any equally definite proof, though arguments of more or less cogency can be elicited from the ancient problem of the One and the Many*. It is not necessary for our purposes, however, to assert dogmatically, that there are no such things as classes. It is only necessary for us to show that the incomplete symbols which we introduce as representatives of classes yield all the propositions for the sake of which classes might be thought essential. (Whitehead and Russell 1913, I, 75–76)

Roughly, the no-class theory proceeds as follows. If ϕx is a matrix (that is, an open formula), we say that $\phi\hat{x}$ is the name of the function involved in this matrix itself. Functions that involve no variables except individuals are called *first-order functions*. Functions which have first-order functions among their arguments and have no arguments except first-order functions and individuals are called second-order. In general, a function is $n+1$-th order ($n > 0$) if it has at least one function of n-th order among its arguments and no function of at least $n + 1$-function among its arguments. A function of order n is called predicative if in its definition only quantification over functions of order $k < n$ is employed. The fact that $\phi\hat{x}$ is predicative is denoted by putting an exclamation mark after ϕ: $\phi!\hat{x}$ and writing correspondingly $\phi!x$ for its value. Two propositional functions are said to be formally equivalent when any argument which satisfies one of them also satisfies the other. A function of a function is called extensional when its truth-value with any argument is the same as with any formally equivalent argument. All functions introduced in the system of PM are extensional.

The authors also accept the Axiom of Reducibility which states that given any function $\phi\hat{x}$ there is a formally equivalent predicative function. For functions of

one and two variables it is formulated (Whitehead and Russell 1913, I, 71–72) as
(I changed the notation slightly):

$$\forall \phi\, \exists \psi\, \forall x\, (\phi x \equiv \psi ! x)$$
$$\forall \phi\, \exists \psi\, \forall x, y\, (\phi(x, y) \equiv \psi !(x, y))$$

Now, every propositional function $\phi \hat{x}$ determines a class $\hat{z}(\phi z)$. If two functions are
formally equivalent, they determine the same class:

$$\forall \phi, \psi\, [\hat{z}(\phi z) = \hat{z}(\psi z) \equiv \forall x\, (\phi x \equiv \psi x)]$$

How do Russell and Whitehead reduce the notion of class which satisfies the
above conditions, but also which allows a name of a class to stand as a grammatical
subject of a sentence to a sentence which does not contain any name of a class? The
three relevant definitions are:

⋆20.01 $f\{\hat{z}(\psi z)\} =_{Df} \exists \phi\, [\forall x\, (\phi ! x \equiv \psi x) \wedge f\{\phi ! \hat{z}\}]$

⋆20.02 $x \in (\phi ! \hat{z}) =_{Df} \phi ! x$

⋆20.03 $cls =_{Df} \hat{\alpha}\{\exists \phi\, (\alpha = \hat{z}(\phi ! z))\}$

The idea is this. If we have a sentence which apparently states something about
a class, we reduce it using **20.01** to a statement about a predicative function (the
existence of which is guaranteed by the Axiom of Reducibility). Specifically, saying
that a certain predicative function determines a class and an object is an element
of this class, by **20.02** boils down to saying that this function applied to this object
yields a true proposition. **20.03** allows us to use the name 'class' itself.

The star in the Leśniewski quotation is a reference to an argument against the
existence of classes put forward by Russell. The argument is:

> If there is such an object as a class, it must be in some sense one object. Yet it is only of
> classes that many can be predicated. Hence, if we admit classes as objects, we must suppose
> that the same object can be both one and many, which seems impossible. (Whitehead and
> Russell 1913, I, 72)

Leśniewski was quite amazed that the argument had been taken seriously (1927, 224–
226). It seemed to him that using the word 'many' makes sense when it is treated as a
quantificational device in the contexts like 'there are many a's' where a is a general
countable term. But then, there is no single object which has the property of being
many. On the other hand, for no object x does it make sense to say that x is many,
because 'many' does not express a property that an object can posses:

> Despite my sincerest wish, I am unable to treat seriously the thesis which proclaims that, "if
> we admit classes as objects, we must suppose that the same object can be both one and many
> which seems impossible", as I feel in it some gross misunderstanding: even posito that the
> expression 'many' causes no doubts on the theme 'at least how many?', I can find no sense
> in saying about some object that it is 'many' even though, by assuming that the meaning
> of the expression 'many' is not uncertain with respect to quantity, that it means e.g., the

same as 'at least two', I fully understand e.g., the utterance that 'many objects' exist in the world ...Seeing no sense in the thesis quoted, I cannot regard it as an even slightly 'cogent' argument for anything at all in the world. (Leśniewski 1927, 225–226)

Leśniewski disagreed also with the no–class theory itself. He believed that the name 'class' has some irreducible use in the natural language. However, he thought it was the mereological use: collections of concrete objects are concrete objects, every class is an element of itself and being an element is the same as being a part (in the improper sense that treats every object as a part of itself). His formalized Mereology was meant to be a theory of classes thus understood.[9]

5.2 Axiomatizations

I will now describe four axiomatizations of Mereology developed by Leśniewski. They come from (respectively) 1916, 1918, 1920, 1921.[10] The first axiomatization was first published in 1928. The axiomatization from 1918 (published in Leśniewski 1930a resulted from Leśniewski's discontent with the fact that the axiomatization from 1916 employed defined terms in two of its axioms. The 1920 axiomatization (published in Leśniewski 1930a as well) took 'being an ingredient' (i.e. a proper or an improper part) as the primitive, instead of 'being a proper part' used in previous axiomatization. His axiomatization from 1921 (published in Leśniewski 1931a) is significantly simpler. It takes 'being exterior' as the primitive and consists of two (as opposed to four) axioms.

A generally interesting question is why Leśniewski waited that long with publication of his results. This question is hard to answer. Most likely, he was a perfectionist and did not want to publish his results before preparing versions which he would consider final. There is only one passage where Leśniewski comments on this issue, and he rather tries to find an excuse for publishing his papers which he does not consider really ready for publication rather than explain why he waited so long. In the beginning of his *On the Foundations of Mathematics* he says:

> The aim of this work is the removal of a painful situation, in which I have found myself for a number of years. The situation consists in this, that I possess a good many unpublished scientific results from various areas of the foundations of mathematics, that the number of these unpublished results continually grows, and since these results inter–relate with

---

[9] Leśniewski also admitted a second use, called 'distributive', which was similar to that formalized by the classical set theory. Even though no published work by Leśniewski contains this claim (he does not even say explicitly that one can make sense of Cantor's set theory or of the system of *Principia* when one considers this to be the sense they were concerned with), it has been made by competent Leśniewskian scholars (Luschei 1962, 29–32; Sobociński 1949b). I will discuss this issue in Chap. 7.

[10] The systems are dated based on Leśniewski's testimony present in the papers in which he describes his systems, which was confirmed by people like Kotarbiński with whom Leśniewski had discussed his results long before he published them (see for instance Kotarbiński 1929). Also, Leśniewski had lectured on his systems while in Warsaw long before they were actually published.

each other and with the results of other researchers in these areas, the technical–editorial difficulties connected with their preparation for publication continually increase.

While I was trying different ways of arranging the scientific results to which I had arrived, I considered among others a systematic–compendium method, I took as a model in this connection the well–known work of Whitehead and Russell. However, such a task is again spreading over a number of years, and it is difficult for me to determine how much time I would still need, in order to submit in this way for a wider technical discussion the whole of the results, to which I have been led by already more than ten years of reflection on the foundation of mathematics.

This painful situation is further complicated by the fact that just as I reached some of my views and some scientific results under the influence of conversations with my colleagues and in connection with their still unpublished scientific results, so also my views and observations, which I had formulated during a number of years in university lectures and in numerous scientific discussions, have contributed to the formation of certain opinions and results of my colleagues, who, out of an admirable loyalty towards me, have withheld from publication until now a number of their scientific results, until my own related results are published. (Leśniewski 1927, 174–175).

Putting these historical issues aside, let us take a look at the axiomatizations themselves.

5.2.1 Axiomatization from 1916

The language in which Mereology was originally formulated was partially informal. However, for the purpose of presentation I will use Ontology as the logical theory which underlies Mereology and I will formalize theorems of Mereology using the language of Ontology. Rules of inference of Mereology are the same as that of Ontology.[11] In the process of formalization it is important to note that 'a is not b' and 'every a is b' for Leśniewski had existential import. They were respectively equivalent to: 'a is an object and not(a is b)' and 'some object is a, and for any X, if X is a, then X is b'. (Leśniewski 1928, 230) Let us start with two preliminary definitions (already given by Leśniewski):

$$\forall a, b \, [a\varepsilon \sim b \equiv a\varepsilon a \land \neg a\varepsilon b]$$
$$\forall a, b \, [a \subseteq b \equiv \forall c \, (c\varepsilon a \rightarrow c\varepsilon b)]$$

The 1916 formulation had two axioms not containing defined terms and two axioms which involved previously defined terms.[12] The first two axioms introduce the notion of a part.

Axiom I $\forall a, b \, (a\varepsilon prt(b) \rightarrow b\varepsilon \sim prt(a))$

that is, if a is a part of b, then b is an object which is not a part of the object a.

---

[11] To avoid indexing variables in the chapter on Mereology I allow all the letters a, b, c, d, e, f, g to be name variables.

[12] Names of formulas in this paragraph come from the 1916 paper.

Axiom II $\forall a, b, c \, (a \varepsilon prt(b) \wedge b \varepsilon prt(c) \rightarrow a \varepsilon prt(c))$

This states the transitivity of parthood relation: if a is a part of b and b is a part of c, then a is a part of c.[13]

Axiom I indicates that strictly speaking by 'part' Leśniewski understood a proper part: a part of an object cannot be identical with that very object. For the reflexive closure of the parthood relation (sometimes called 'improper parthood') he introduced the term 'ingredient',[14] defined by:

Definition I $\forall a, b \, (a \varepsilon ingr(b) \equiv (a \varepsilon b \wedge b \varepsilon a) \vee a \varepsilon prt(b))$

which reads: a is an ingredient of b if and only if either a and b are the same object, or a is a part of b. Two other important definitions are that of a set and that of a class. As it will turn out, this has nothing to do with the standard set-theoretic distinction between sets and classes. An object a is a set of b (or a set of bs) if and only if it is an object such that every one of its ingredients has an ingredient which is an ingredient of an object which is b:

Definition II $\forall a, b \, (a \varepsilon set(b) \equiv a \varepsilon a \wedge \forall c \, (c \varepsilon ingr(a) \rightarrow$
$\rightarrow \exists d \, (d \varepsilon ingr(c) \wedge \exists e \, (e \varepsilon b \wedge d \varepsilon ingr(e) \wedge e \varepsilon ingr(a)))))$

For instance, the mereological whole constituted, say, by all people in Canada is a set of people, because every one of its ingredients has an ingredient which is an ingredient of a person (in Canada).

To see why this rather complicated iteration of "ingredienthood" is needed, consider what would happen if the definition required only that every ingredient of a be an ingredient of an object which is b. Then, the mereological whole constituted by all people in Canada (call it CAN) would not be a set of people, because CAN would have ingredients, like the mereological fusion of one person's leg and another person's right hand, which would not be ingredients of any particular person.

Also, note that the indefinite article in 'is a set' is there not without a purpose. According to this definition, one countable noun may generate quite a few different sets. The intuition here is that every mereological whole constituted by **some** objects b (i.e. by some objects denoted by the countable noun phrase 'b') is a set of b, although choosing different representatives (or groups of representatives) of b we get different mereological wholes.

---

[13] A quick reminder about the notation. I abbreviated Leśniewski's wording of formulas. Instead of 'for some' I write '∃', instead of 'for all' I write '∀'. Instead of 'is' I write 'ε'. I also sometimes changed the shape of brackets. Also, informally, Leśniewski used capital variables in case when the truth of a formula required that a term be singular. Otherwise, the formulas are pretty much the same formulas that Leśniewski gave. So, for instance, originally, Axiom II would read: 'If P is a part of object Q, and Q is a part of object R, then P is a part of object R.' It is important to remember that he did not have two sorts of variables in his official language, and only used capital letters as variables in formulas for whose satisfaction it was required that those variables be interpreted as individual terms. To avoid shaping syntax by semantics and to keep things simple I decided to use only lower case variables. If the reader is not satisfied with my choice, he is free to rewrite all relevant formulas in his favorite notation.

[14] The term is claimed to have been suggested by Mr Lucjan Zarzecki (1928, 230).

Take another example: any heap of stones is a set of stones in the sense introduced by Definition II. Therefore, if a names more than one object, the name $set(a)$ names also more than one object: it names any mereological whole constructed from some objects that fall under a.

Another notion defined by Leśniewski is the one that determines the 'maximal' set of objects a, that is the set of all a's. This is the notion of a class, and an object a is a class of objects b if and only if every object b is an ingredient of a, and any ingredient of a has an ingredient which is also an ingredient of some object b:[15]

Definition III $\forall a, b \, (a\varepsilon cl(b) \equiv a\varepsilon a \wedge \forall c \, (c\varepsilon b \rightarrow c\varepsilon ingr(a)) \wedge$
$\wedge \forall c \, (c\varepsilon ingr(a) \rightarrow \exists d \, (d\varepsilon ingr(c) \wedge \exists e \, (e\varepsilon b \wedge d\varepsilon ingr(e)))))$

The remaining two axioms employ the notion of a class:

Axiom III $\forall b \, (\exists a \, a\varepsilon b \rightarrow \exists c \, c\varepsilon cl(b))$

Axiom III may be meta-linguistically read as stating that any non-empty name generates its class.

Axiom IV $\forall a, b, c \, (a\varepsilon cl(b) \wedge c\varepsilon cl(b) \rightarrow a\varepsilon c)$

Axiom IV states that a given name generates at most one class.

5.2.2 Axiomatization from 1918

Quite soon Leśniewski became dissatisfied with the fact that his formulation employed defined terms in two of its axioms.[16] Thus, in 1918 he made a modification described in his paper (Leśniewski 1930a). Definitions I, II and III and axioms I and II remain untouched. Axioms III and IV are replaced by:

Axiom C $\forall a, b, d \, [b\varepsilon b \wedge d\varepsilon d \rightarrow \forall c \, (c\varepsilon a \rightarrow ((c\varepsilon b \wedge b\varepsilon c) \vee c\varepsilon prt(b))) \wedge$
$\wedge \forall c \, (c\varepsilon a \rightarrow ((c\varepsilon d \wedge d\varepsilon c) \vee c\varepsilon prt(d))) \wedge \forall e \, (e\varepsilon prt(b) \vee e\varepsilon prt(d) \rightarrow$
$\rightarrow \exists f \, (((f\varepsilon e \wedge e\varepsilon f) \vee f\varepsilon prt(d)) \wedge (f\varepsilon a \vee \exists c \, (c\varepsilon a \wedge f\varepsilon prt(c))))) \rightarrow b\varepsilon d]$

Axiom D $\forall a \, [\exists c \, c\varepsilon a \rightarrow \exists b \, (\forall d \, (d\varepsilon a \rightarrow (d\varepsilon b \wedge b\varepsilon d) \vee d\varepsilon prt(b)) \wedge$
$\wedge \forall d \, (d\varepsilon prt(b) \rightarrow \exists c \, (((c\varepsilon d \wedge d\varepsilon c) \vee c\varepsilon prt(d)) \wedge (c\varepsilon a \vee \exists e \, (e\varepsilon a \wedge c\varepsilon prt(e))))))]$

Axiom C looks somewhat complicated, but its intuitive sense is not so difficult. Less formally speaking, it says that if:

---

[15] Since $cl(a)$ is the mereological fusion of all those objects that are a, I will use 'the class of a' and 'the class of a's' interchangeably. By the former use I want to emphasize that it is a class determined by a single name a and by the latter I want to highlight that it is nothing above and beyond the objects that a already denotes, considered as one individual object.

[16] Keep in mind that making your axiomatization as simple as possible was the name of the game in Leśniewski's days. People not only wanted to have as few axioms as possible, but they also wanted their axioms to contain as few symbols as possible. One of the mysteries in the history of logic is why anyone actually cared about things like replacing an axiom with another axiom, say, two symbols shorter, and proving they were equivalent.

1. b and d are objects ($b \varepsilon b \wedge d \varepsilon d$):
2. any object c which is a is an ingredient of b:

$$\forall c \, (c \varepsilon a \rightarrow ((c \varepsilon b \wedge b \varepsilon c) \vee c \varepsilon prt(d)))$$

(by Definition I, c is an ingredient of b iff either it is identical to b ($c \varepsilon b \wedge b \varepsilon c$), or it is a part of d.)
3. any a is an ingredient of d:

$$\forall c \, (c \varepsilon a \rightarrow ((c \varepsilon d \wedge d \varepsilon c) \vee c \varepsilon prt(d)))$$

4. any part of either b or d:

$$\forall e \, (e \varepsilon prt(b) \vee e \varepsilon prt(d) \rightarrow$$

has an ingredient:
$$\exists f \, (((f \varepsilon e \wedge e \varepsilon f) \vee f \varepsilon prt(d)) \wedge$$

which either is a
$$(f \varepsilon a \vee$$

or is a part of one of a's:
$$\exists c \, (c \varepsilon a \wedge f \varepsilon prt(c)))$$

then b is d ($b \varepsilon d$).

So, the axiom says that if b and d are two classes (in Leśniewski's sense) of a's, then they are identical (in fact, this is pretty much a restatement of Axiom IV, but without the use of defined terms).

Axiom D (when we introduce the notion of ingredient using definition I) may be read as stating that if at least one object is a, then there is an object such that all objects a are its ingredients and every part of it has an ingredient which is an a or an ingredient of an a. In other words: if at least one object is a, then there is a class of objects a.

This formulation is not deeply innovative. It is easy to see that axiom D results from axiom III if we eliminate from it the notions of ingredient and class using definitions I and III. For Axiom III says that for any b, if there is an a such that $a \varepsilon b$ (i.e. b is non–empty), there is an object which is a class of b's (I used the indefinite article because the axiom does not imply that such a class is unique). When we use Definition III, the consequent of Axiom III reads:

$$\exists c \, [c \varepsilon c \wedge \forall d \, (d \varepsilon b \rightarrow d \varepsilon ingr(c)) \wedge \forall d \, (d \varepsilon ingr(c) \rightarrow \exists e \, (e \varepsilon ingr(d) \wedge$$
$$\wedge \exists f \, (f \varepsilon b \wedge e \varepsilon ingr(f))))]$$

Now, using Definition I we replace every expression of the form $a\varepsilon ingr(\beta)$ with $(a\varepsilon\beta \wedge \beta\varepsilon a) \vee a\varepsilon prt(\beta)$, and we get a formula which pretty much is Axiom D. The only difference is that it contains $c\varepsilon c$ in the consequent. This, however, is logically redundant, because $\exists a\, a\varepsilon b$ together with $\forall d\, (d\varepsilon b \rightarrow d\varepsilon ingr(c))$ already imply $\exists d\, d\varepsilon ingr(c)$ which can be true only if c is a singular term, that is only if $c\varepsilon c$.

5.2.3 Axiomatization from 1920

In (1930a) Leśniewski discussed a formulation of Mereology dating from 1920 which takes 'is an ingredient' as a primitive term. This formulation is equivalent to the previous formulations. It has four axioms:

Axiom (a) $\forall a, b\, (a\varepsilon ingr(b) \wedge \neg b\varepsilon a \rightarrow b\varepsilon \sim a)$

Axiom (b) $\forall a, b, c\, (a\varepsilon ingr(b) \wedge b\varepsilon ingr(c) \rightarrow a\varepsilon ingr(c))$

Axiom (c) $\forall a, c, d\, [\forall b\, (b\varepsilon a \rightarrow b\varepsilon ingr(c) \wedge b\varepsilon ingr(d)) \wedge$
$\wedge \forall b\, (b\varepsilon ingr(c) \vee b\varepsilon ingr(d) \rightarrow \exists e\, (e\varepsilon ingr(b) \wedge$
$\wedge \exists f, g\, (f\varepsilon ingr(g) \wedge g\varepsilon a \wedge e\varepsilon ingr(f)))) \rightarrow c\varepsilon d]$

Axiom (d) $\forall a\, [\exists c\, c\varepsilon a \rightarrow \exists b\, (\forall d\, (d\varepsilon a \rightarrow d\varepsilon ingr(b)) \wedge$
$\wedge \forall d\, (d\varepsilon ingr(b) \rightarrow \exists c\, (c\varepsilon ingr(d) \wedge \exists e\, (e\varepsilon a \wedge c\varepsilon ingr(e)))))]$

and parthood is defined by:

Definition (e) $\forall a, b\, (a\varepsilon prt(b) \equiv a\varepsilon ingr(b) \wedge \neg(a\varepsilon b \wedge b\varepsilon a))$

The definition of class is the same as Definition III in the version from 1916.

Again, there seems to be nothing essentially new about this formulation. It takes another expression as a primitive, but the reformulation of its axioms is a quite straightforward translation of axioms occurring in previous formulations. And so, Axiom (a), given Definition (e), has the same intuitive content as Axiom I. Axiom (b), considering the same definition easily implies Axiom II. Indeed, Axiom (b) seems just to be 'a reflexive closure' of the assumption of transitivity from axiom II. Axiom (c) is just another formulation of the extensionality statement, present already in Axiom IV. Axiom (d), if we use definition III turns out to boil down to Axiom IV.

5.2.4 Axiomatization from 1921

In (1931a) Leśniewski discusses another, equivalent formulation which takes 'exterior to' as a primitive. Two objects are exterior to each other if and only if they have no overlapping parts.[17] There are two axioms:

---

[17] What he published in 1931 is nevertheless referred to as the 1921 axiomatization, since this seems to be the year when it was formulated. This difference arose from the fact that Leśniewski lingered with the publication of his results.

Axiom $\mathfrak{A}$ $\forall a, b\,[a\varepsilon ext(b) \equiv \forall c\,(c\varepsilon c \to$
$\to \exists d\,((d\varepsilon ext(a) \vee d\varepsilon ext(b)) \wedge d\varepsilon \sim ext(c)))]$

Axiom $\mathfrak{B}$ $\forall a, \phi\,[\forall b\,[b\varepsilon\phi(a) \equiv b\varepsilon b \wedge \exists d\,d\varepsilon a \wedge$
$\wedge \forall c\,(b\varepsilon ext(c) \equiv \forall d\,(d\varepsilon a \to d\varepsilon ext(c)))] \to$
$\to \phi(a)\varepsilon\phi(a)]$

Axiom $\mathfrak{A}$ can be read as saying that two objects, a and b are exterior to each other if and only if for any object c there is an object d which overlaps with c but which does not overlap with either a or b. To grasp intuitively why the equivalence holds, consider the following. From right to left: suppose a and b are not exterior to each other. This means that they have a common part. Call it c. But then, there is no d overlapping with c which does not overlap with a and b. For the opposite direction, suppose there is a c such that any d which overlaps with c, overlaps both with a and b. This means that c is a common ingredient of a and b and hence, a and b overlap.

Axiom $\mathfrak{B}$ may be taken to state that if something is a, then there is a class of a's. Basically what it says is this: suppose you define a name–forming functor of one name argument, ϕ, by saying that it names any object b such that (a) if an object c is exterior to all the a's, b is exterior to c, and (b) if b is exterior to an object c, then c is exterior to all the a's. Part (a) means that b has no ingredient outside of the fusion of the a's. Part (b) means that there is no part of the fusion of all the a's which is exterior to b. Together, they mean that b is the maximal object which has all a's as its ingredients. That is, b is their fusion. So, given that '$\phi(a)$' is defined as 'the fusion of a's', Axiom $\mathfrak{B}$ says that if a itself is non–empty then there is exactly one fusion of a's ('$\phi(a)\varepsilon\phi(a)$' is true only if there is exactly one object denoted by '$\phi(a)$').

The notions of a class and of an ingredient can be easily defined:

Definition $\mathfrak{C}$ $\forall a, b\,[b\varepsilon cl(a) \equiv b\varepsilon b \wedge \forall c\,(b\varepsilon ext(c) \equiv \forall d\,(d\varepsilon a \to d\varepsilon ext(c)))]$

This reads: b is a class of a's iff b is an object and it is exterior to an object c if and only if this object is exterior to all the a's. It is easy to see that this definiens occurs in Axiom $\mathfrak{B}$.

Definition $\mathfrak{D}$ $\forall a, b\,[a\varepsilon ingr(b) \equiv \exists c\,(b\varepsilon cl(c) \wedge a\varepsilon c)]$

This definition says that an object a is an ingredient of an object b if and only if for some c, b is a class of c's and a is one of the c's. The parthood relation is defined as in Definition (e).

This axiomatization interestingly diverges from the previous ones and the proof of their equivalence (given by Leśniewski in the same paper) is rather non-trivial.

5.2.5 Some Other Conditions of Being a Class

K. Kuratowski and A. Tarski formulated three other possible definitions of class, equivalent to the one used originally by Leśniewski. These results are described in Leśniewski (1930a).

Kuratowski's definition is:

Definition a $\forall a, b \,[b\varepsilon cl(a) \equiv \forall c\,(c\varepsilon a \rightarrow c\varepsilon ingr(b)) \wedge$
$\wedge \forall c\,(\forall d\,(d\varepsilon a \rightarrow d\varepsilon ingr(c)) \wedge c\varepsilon ingr(b) \rightarrow b\varepsilon c \wedge c\varepsilon b)]$

Roughly speaking, it says that an object b is $cl(a)$ if (1) it mereologically contains all the a's, and (2) whatever mereologically contains all a's and is an ingredient of b simply is b. So, for instance if you take a clay statue, call it 'Jay', Jay will be the class of 'all clay pieces constituting the statue' since whatever is such a clay piece is an ingredient of Jay, and any object which mereologically contains the fusion of all such clay pieces and itself is an ingredient of Jay simply is Jay (this suggests that plain classical mereology is not the best framework for the constitution-identity debate).

Two definitions suggested by Tarski are:

Definition b $\forall a, b \,[b\varepsilon cl(a) \equiv \forall c\,(c\varepsilon a \rightarrow c\varepsilon ingr(b)) \wedge$
$\wedge \forall c\,(\forall d\,(d\varepsilon a \rightarrow d\varepsilon ingr(c)) \rightarrow b\varepsilon ingr(c)))]$

Definition c $\forall a, b \,[b\varepsilon cl(a) \equiv b\varepsilon b \wedge \forall c\,(\forall d\,(d\varepsilon a \rightarrow$
$\rightarrow d\varepsilon ingr(c)) \equiv b\varepsilon ingr(c))]$

Tarski's **Definition b** is slightly different from Kuratowski's. It says that b is a class of a's if it mereologically contains all a's, and is mereologically contained in any object that mereologically contains all a's.

Definition c on the other hand says that for b to be a class of a's it is enough that (1) b is an object, that (2) it is an ingredient of anything that contains mereologically all a's, and that (3) if it is an ingredient of an object c, this object mereologically contains all a's.

5.3 Mereology as a Replacement for Set Theory

Leśniewski intended to develop Mereology as an alternative to classical set theory. In the series of papers in which he described various axiomatizations of Mereology he also proved numerous theorems that are intended as analogues of certain theorems of set theory. Although his proofs are not wholly formalized, they are clear enough to see that they can be fully formalized. Also, an interesting aspect of Leśniewski's proofs in Mereology is that he clearly uses a strategy typical for natural deduction (freezing variables, assuming the antecedent and proving the consequent in order to prove the whole conditional), even though the official deductive system developed by him does not allow for this kind of moves. Nowadays it is a common practice, but those days it stood in clear contrast with the more common strategy of proving theorems within a system in the Hilbert-style fashion.[18]

---

[18] It is said that Leśniewski eventually became dissatisfied with Mereology as a system of foundations of mathematics. He did not publish such a comment anywhere, and the information seems to come from secondary sources. For instance (Kearns 1962, 35) says: "Leśniewski did not give up

I will now give and describe those theorems of Mereology proven by Leśniewski which are crucial for one of his stronger theorems in Mereology, an analogue of Cantor's theorem. (Almost all theorems I will describe have been given and proven by Leśniewski. Occasionally I will also mention fairly trivial consequences of his theorems.)

This analogue states that if a name a names more than one object, and the objects it names do not overlap, then there are more sets of a's than objects a's. So, for instance, if I have five (obviously, non overlapping) lizards, and I name this little herd 'Aldos', then the theorem implies that there are more than five sets of Aldos.

Going over a few theorems on our way will also help us understand the expressive power of the language of Mereology. Simultaneously I will also discuss those theorems which show how Mereology differs from set theory. For the sake of accessibility I replace proofs given by Leśniewski with informal explanations. The reader is free to look at detailed proofs in Leśniewski's own writings. (Those who are eager to see some examples of proofs, I beg to wait till Chap. 7.) Numbering of the following theorems and definitions has nothing to do with the names assigned by Leśniewski, but in footnotes I give detailed references to the original formulations. For theorems proven by Leśniewski I provide references in footnotes.

Among some simple theorems which involve the notions of parthood and ingredient are:

$$\forall a \, (a\varepsilon a \rightarrow a\varepsilon \neg prt(a)) \tag{5.1}$$

which states that no object is part of itself.[19] It easily follows from Axiom I.

$$\forall a \, (a\varepsilon a \rightarrow a\varepsilon ingr(a)) \tag{5.2}$$

that is, every object is an ingredient of itself.[20] Definition I implies:[21]

$$\forall a, b \, (a\varepsilon prt(b) \rightarrow a\varepsilon ingr(b)) \tag{5.3}$$

which means that every part of an object is also its ingredient. Also, the relation of being an ingredient is transitive:[22]

$$\forall a, b, c \, (a\varepsilon ingr(b) \wedge b\varepsilon ingr(c) \rightarrow a\varepsilon ingr(c)) \tag{5.4}$$

---

his claims that it [Mereology] is intuitively acceptable, but he did realize that it cannot replace set theory."

[19] This is Theorem I from Leśniewski (1916, 131). Also proven as Theorem I in Leśniewski (1928, 232).

[20] Theorem II, (Leśniewski 1916, 132). Also, Theorem II in Leśniewski (1928, 233).

[21] Theorem III, (Leśniewski 1916, 132).

[22] Theorem IV, (Leśniewski 1916, 132). Theorem IV in Leśniewski (1928, 234).

If one object is an ingredient of another object, they have at least one ingredient in common:[23]

$$\forall a, b \, (a\varepsilon ingr(b) \rightarrow \exists c \, (c\varepsilon ingr(a) \wedge c\varepsilon ingr(b))) \tag{5.5}$$

If an object a is an ingredient of another object b, then a has an ingredient which is an ingredient of an ingredient of b:[24]

$$\forall a, b \, (a\varepsilon ingr(b) \rightarrow \exists d \, (d\varepsilon ingr(b) \wedge \exists c \, (c\varepsilon ingr(a) \wedge c\varepsilon ingr(d)))) \tag{5.6}$$

Given the Leśniewskian understanding of the notions of class and set, clearly every class is a set:[25]

$$\forall a, b \, (a\varepsilon cl(b) \rightarrow a\varepsilon set(b)) \tag{5.7}$$

Every object is the class of its own ingredients:[26]

$$\forall a \, (a\varepsilon a \rightarrow a\varepsilon cl(ingr(a))) \tag{5.8}$$

and whatever has a part is the class of its own parts:[27]

$$\forall b \, (\exists a \, a\varepsilon prt(b) \rightarrow b\varepsilon cl(prt(a))) \tag{5.9}$$

The identification of a singleton with its element (mentioned before) is also provable:[28]

$$(a\varepsilon a \rightarrow a\varepsilon cl(a)) \tag{5.10}$$

The definition of being an element abides by the intuition that for a to be an element of a class b there has to be some c, such that b is the class of c, and a is c:

Def. (el) $\forall a, b \, [a\varepsilon el(b) \equiv \exists c \, (b\varepsilon cl(c) \wedge a\varepsilon c)]$

Interestingly, if a is an ingredient of b, then by (5.8) there is a name $ingr(b)$ such that $b\varepsilon cl(ingr(b))$ and $a\varepsilon ingr(b)$, and thus a is an element of b:[29]

$$\forall a, b \, (a\varepsilon ingr(b) \rightarrow a\varepsilon el(b)) \tag{5.11}$$

Also, if a is an element of b, then there is some c such that $b\varepsilon cl(c)$ and $a\varepsilon b$, and thus a is an ingredient of b (for, by definition III, if b is $cl(c)$, then every c is its ingredient):[30]

---

[23] Theorem V, (Leśniewski 1916, 133).

[24] Theorem VI, (Leśniewski 1916, 134).

[25] Theorem VII, (Leśniewski 1916, 136).

[26] Theorem VIII, (Leśniewski 1916, 136). Also, Theorem VII, (Leśniewski 1928, 236).

[27] Theorem IX, (Leśniewski 1916, 136–137).

[28] Theorem X, (Leśniewski 1916, 138).

[29] Theorem XI, (Leśniewski 1916, 139).

[30] Theorem XII, (Leśniewski 1916, 139).

$$\forall a, b \, [a\varepsilon el(b) \rightarrow a\varepsilon ingr(b)] \tag{5.12}$$

Theorems (5.11) and (5.12) together identify being an element of a class with being its ingredient. This identification implies the reflexivity and transitivity of being an element:[31]

$$\forall a \, (a\varepsilon a \rightarrow a\varepsilon el(a)) \tag{5.13}$$

$$\forall a, b, c \, (a\varepsilon el(b) \land b\varepsilon el(c) \rightarrow a\varepsilon el(c)) \tag{5.14}$$

The class of sets of b is the class of b's:[32]

$$\forall a, b \, (a\varepsilon cl(set(b)) \rightarrow a\varepsilon cl(b)) \tag{5.15}$$

Any set of b is an ingredient of the class of b's:[33]

$$\forall a, b \, (a\varepsilon set(b) \rightarrow a\varepsilon ingr(cl(b))) \tag{5.16}$$

Any class of b's is a class of sets of b's:[34]

$$\forall a, b \, (a\varepsilon cl(b) \rightarrow a\varepsilon cl(set(b))) \tag{5.17}$$

From (5.7) and the identification of being an ingredient with being an element it follows that every object is the class of its elements:[35]

$$\forall a \, (a\varepsilon a \rightarrow a\varepsilon cl(el(a))) \tag{5.18}$$

Theorem (5.13) implies that every object is an element of itself. Note also that for any a there is at most one class of a's:

$$\forall a, b, c \, (a\varepsilon cl(c) \land b\varepsilon cl(c) \rightarrow a\varepsilon b) \tag{5.19}$$

Now, if a is a class of b's, it is unique. Thus, being an object, by (5.13) it is an element of itself. That is, every class has at least one element: itself.[36]

$$\forall a, b \, (a\varepsilon cl(b) \rightarrow a\varepsilon el(a)) \tag{5.20}$$

Therefore, there is no empty class.

---

[31] Theorems XIV and XV, (Leśniewski 1916, 140).
[32] Theorem XX, (Leśniewski 1916, 143).
[33] Theorem XXI, (Leśniewski 1916, 145).
[34] Theorem XXII, (Leśniewski 1916, 145).
[35] Theorem XXIV, (Leśniewski 1916, 150).
[36] Theorem XXV, (Leśniewski 1916, 150).

$$\neg \exists a \, \forall b \, \neg b \varepsilon el(a) \tag{5.21}$$

One can define a name: 'a class which is not an element of itself':

Def. (r) $\forall a \, [a\varepsilon \underline{r} \equiv a\varepsilon a \wedge \neg a\varepsilon el(a) \wedge \exists b \, a\varepsilon cl(b)]$

The class of classes which are not elements of themselves would be $cl(\underline{r})$. But, since every class is an element of itself and no class is empty, $cl(\underline{r})$ does not exist:[37]

$$\neg \exists a \, a\varepsilon cl(\underline{r}) \tag{5.22}$$

The notions of subset and proper subset are defined as follows:

Def. (subset) $\forall a, b \, (a\varepsilon subset(b) \equiv a\varepsilon a \wedge \forall c \, (c\varepsilon el(a) \rightarrow c\varepsilon el(b)))$

Def. (psubset) $\forall a, b \, (a\varepsilon psubset(b) \equiv a\varepsilon subset(b) \wedge \neg b\varepsilon a)$

Those notions are not very fascinating, though:[38]

$$\forall a, b \, (a\varepsilon subset(b) \equiv a\varepsilon ingr(b)) \tag{5.23}$$

$$\forall a, b \, (a\varepsilon psubset(b) \equiv a\varepsilon prt(b)) \tag{5.24}$$

$$\forall a, b \, (a\varepsilon subset(b) \equiv a\varepsilon el(b)) \tag{5.25}$$

It is also possible to define 'the universe'. First, we have to define the name 'object':

Def. (obj) $\forall a \, (a\varepsilon obj \equiv a\varepsilon a)$

and now we say that the universe is the class of all objects:[39]

Def. (U) $\forall a \, (a\varepsilon U \equiv a\varepsilon a \wedge a\varepsilon cl(obj))$

In the first paper on Mereology (Leśniewski 1916) there is an interesting confusion which shows that Leśniewski was not clear as to the logical foundations of his system. Theorem XLIII (Leśniewski 1916, 159) claims that some object is the class of non-contradictory objects. The informal proof is:

> In agreement with the law of non-contradiction we may state that every object is a non-contradictory object. It follows from this that some object is a non-contradictory object, from which—in accordance with Axiom III—the given theorem results. (Leśniewski 1916, 159)

Theorem XLIV (p. 160) states that the class of non-contradictory objects is the universe.

---

[37] A slight reformulation of Theorem XXVI, (Leśniewski 1916, 150).

[38] These are obvious consequences of Theorems XXVIII–XXXI, (Leśniewski 1916, 152–155).

[39] Definition VII, (Leśniewski 1916, 159).

Neither XLIII nor XLIV actually are theorems of Mereology. They both imply that there exists at least one object. However, in the empty domain all axioms of Ontology and Mereology are valid and all rules are truth–preserving (or, in other words, the axiomatic basis of Leśniewski's systems does not contain unconditional existential statements).[40] So, the alleged Theorems XLIII and XLIV imply something that Mereology does not imply.

In formulations other than that from 1921 the notion of being exterior can be defined as follows:[41]

Def. (ext) $\quad \forall a, b \, (a \varepsilon ext(b) \equiv a \varepsilon a \wedge \forall c \, (c \varepsilon ingr(b) \rightarrow \neg c \varepsilon ingr(a)))$

Thus, a is exterior to b iff a is an object which has no common ingredients with b.

The notion of a relative complement is also definable. The idea is that a is a complement of b respective to c (that is, a is c "minus" b) iff b is a part of c and a is the class of those parts of c which are exterior to b:[42]

Def. (compl) $\quad \forall a, b, c \, [a \varepsilon compl(b, c) \equiv b \varepsilon subset(c) \wedge \exists d \, (a \varepsilon cl(d) \wedge$
$\wedge \forall e \, (e \varepsilon d \equiv e \varepsilon el(c) \wedge e \varepsilon ext(b)))]$

Clearly, if a is a part of b, then there is a's complement relative to b:[43]

$$\forall a, b \, (a \varepsilon prt(b) \rightarrow \exists c \, (c \varepsilon compl(a, b))) \tag{5.26}$$

Some fairly simple theorems which relate the notions of parthood and complement are:[44]

$$\forall a, b, c \, (a \varepsilon compl(b, c) \rightarrow a \varepsilon ext(b)) \tag{5.27}$$

$$\forall a, b, c \, (a \varepsilon compl(b, c) \rightarrow a \varepsilon prt(c)) \tag{5.28}$$

$$\forall a, b, c \, (a \varepsilon compl(b, c) \rightarrow b \varepsilon prt(c)) \tag{5.29}$$

Also, no object is a complement of itself with respect to any class, neither is it the complement of a class with respect to itself, and no class is the complement of a class with respect to that class:[45]

$$\neg \exists a, b \, [a \varepsilon compl(a, b)] \tag{5.30}$$

$$\neg \exists a, b \, [a \varepsilon compl(b, a)] \tag{5.31}$$

$$\neg \exists a, b \, [a \varepsilon compl(b, b)] \tag{5.32}$$

---

[40] See (Simons 1995) for a proof.
[41] Definition VIII, (Leśniewski 1916, 161).
[42] Definition IX, (Leśniewski 1916, 162).
[43] Theorem XLVIII, (Leśniewski 1916, 163).
[44] Theorems IL–LII, (Leśniewski 1916, 164–170).
[45] Theorems LV–LVII, (Leśniewski 1916, 170–173).

Another interesting notion is the notion of a sum of two exterior objects:[46]

Def. (+) $\forall a, b, c \, [a \varepsilon b + c \equiv a \varepsilon cl(b \cup c) \wedge b \varepsilon ext(c)]$

where $\forall a, b, c \, (a \varepsilon b \cup c \equiv a \varepsilon b \vee a \varepsilon c)$. That is, a is the sum of two exterior objects b and c iff a is the class of things which are b or c. It is provable that for any a and b at most one object is $a + b$,[47] and that if a is exterior to b, there exists an object which is $a + b$:[48]

$$\forall a, b, c, d \, (a \varepsilon b + c \wedge d \varepsilon b + c \rightarrow a \varepsilon d) \tag{5.33}$$

$$\forall a, b \, (a \varepsilon ext(b) \rightarrow \exists c \, c \varepsilon a + b) \tag{5.34}$$

* A generalization of $+$ is the notion of sum:[49]

Def. (sum) $\forall a, b \, [a \varepsilon sum(b) \equiv a \varepsilon cl(b) \wedge \forall c, d \, (c \varepsilon b \wedge d \varepsilon b \rightarrow$
$\rightarrow c \varepsilon d \vee c \varepsilon ext(d))]$

Say b is a name which names only pairwise exterior objects. Then the sum of b's is the class of b's.

Probably the most interesting theorem of Mereology given by Leśniewski is the one that corresponds to Cantor's theorem in classical set theory. To be able to formulate it, we first have to define two Ontological concepts: equinumerosity and 'being fewer than'.[50]

First, objects a are as numerous as objects b ($a \simeq b$) iff there is some constant τ of category $\frac{n}{n}$ such that all the conditions below hold:

$$\forall c \, (c \varepsilon a \rightarrow \exists d \, (d \varepsilon b \wedge d \varepsilon \tau(c)))$$
$$\forall d \, (d \varepsilon b \rightarrow \exists c \, (c \varepsilon a \wedge d \varepsilon \tau(c)))$$
$$\forall c, d, e \, (c \varepsilon a \wedge d \varepsilon b \wedge e \varepsilon b \wedge d \varepsilon \tau(c) \wedge e \varepsilon \tau(c) \rightarrow d \varepsilon e \wedge e \varepsilon d)$$
$$\forall c, d, e \, (c \varepsilon a \wedge d \varepsilon a \wedge e \varepsilon b \wedge e \varepsilon \tau(c) \wedge e \varepsilon \tau(d) \rightarrow c \varepsilon d \wedge d \varepsilon c)$$

It is not too difficult to notice that the first condition says that τ relates each object which is a to an object which is b (that is, that τ is total on a's), the second condition says that each object which is b is the "value" of τ for some object which is a (that τ is onto b), the third condition says that τ is functional (that it doesn't relate an a-object to more than one b-object), and the last condition requires that τ should be an injection.

The second defined notion is that of objects a being fewer than objects b. There are fewer a's than b's iff the following two conditions are both satisfied:

---

[46] Definition VIII, (Leśniewski 1929b, 295).

[47] Theorem CLIII, (Leśniewski 1929b, 296).

[48] Theorem CLV, (Leśniewski 1929b, 297).

[49] Definition IX, (Leśniewski 1929b, 302).

[50] Those are explained in footnote 14, (Leśniewski 1929b, 311).

$$\exists c \, [\forall d \, (d\varepsilon c \rightarrow d\varepsilon b) \wedge a \simeq c]$$
$$\forall c \, [\forall d \, (d\varepsilon c \rightarrow d\varepsilon a) \rightarrow \neg b \simeq c]$$

The first requirement is that a is equinumerous to a certain name c which is weakly included in b. The second is that b is not equinumerous to any name c which is weakly included in a.

If we write 'objects a are fewer than objects b' as '$a < b$', the "Cantorian" theorem in question is:[51]

$$\forall a \, [\forall b, c \, (b\varepsilon a \wedge c\varepsilon a \rightarrow c\varepsilon b \vee c\varepsilon ext(b)) \wedge$$
$$\wedge \exists b, c \, (b\varepsilon a \wedge c\varepsilon a \wedge \neg b\varepsilon c) \rightarrow a < set(a)]$$

What this states may intuitively be grasped as follows. Let a be a 'discrete' name, that is, if it names more than one object, those objects have no parts in common. If, moreover, the name names more than one object, then there are more sets of a's than objects a. For example, suppose that 'a' names exactly two stones: b and c, exterior to each other. How many objects a are there? Two. b and c. How many sets of a are there? At least three: b itself, c itself and the sum of b and c are all sets of a. Is, say, a half of b a set of a's? No. According to definition II, any set of a's has to have at least one object a as its ingredient. How about the sum of b and a half of c? If it was a set of a's, then every ingredient of it would have to have one ingredient e such that e would have been an ingredient of an ingredient of one of the a's, which would also have to be an ingredient of this sum itself. But the half of c is not an ingredient of an ingredient of any object a which is an ingredient of the sum in question. So, it seems, there are exactly three sets of a's and only two objects a.

Mereology was constructed by Leśniewski, as he explicitly stated, to provide an alternative form of the foundation of mathematics. This new system indeed provides an interesting theory of classes, at least in one sense of this word.

Unfortunately, neither Leśniewski nor later developments have shown that Mereology actually constitutes a foundation of mathematics in any sense remotely similar to that in which classical set theory does. The last theorem of this section is probably the most mathematically interesting theorem of Mereology that Leśniewski has proven. On the other hand, some quasi-mathematical concepts, like equinumerosity, are already definable in Ontology. Leśniewskian scholars sometimes put it by saying that number is an Ontological concept.

What exactly would Mereology have to be like to succeed as a system of foundations of mathematics, or at least as a replacement for set theory? Considering the importance of classical set theory nowadays I would conjecture that Mereology would be successful if (a) a theory isomorphic with one of standard set theories were a sub–theory of Mereology (or of a nominalistically acceptable extension of it), and yet (b) Mereology would not be committed to abstract objects. Restriction (b) results from Leśniewski's nominalistic assumptions. Restriction (a) is not exactly

---

[51] Theorem CXCVIII, (Leśniewski 1929b, 313).

Leśniewskian. Of course, from what we know, he attempted to prove mereological analogues of certain theorems of the standard set theory. So, if we wanted to be more in accordance with Leśniewski's intuitions, we should require that Mereology be able to prove analogues of major theorems of set theory. The problem with this formulation is that the notion of a mereological analogue of a major set–theoretic theorem is fairly vague. I would suggest (a) as a nowadays reasonable requirement for quite simple pragmatical reasons. There have been thousands of mathematicians who believe set theory makes sense and who use set theory in their everyday research practice. Of course, one can go ahead and say that this theory simply does not make sense and people who are using it are all mistaken, but this would be quite a strong claim. Rather, I am inclined to believe, considering the abundance of set theoretic content in various meta–logical and mathematical contexts, one of the tasks of a person who constructs a foundational system is to explain the meaning of the theory, not to try to explain the theory away. Hence, if one starts with saying that, for instance, it does not even make sense to use the expression 'the empty set' instead of trying to explain how it is possible that this expression makes sense, one puts forward a very revolutionary approach to the problems.

Such a revolutionary strategy might be suspected of not being a foundations of mathematics itself, but rather an attempt to construct a new mathematical theory. The issue here resembles the qualms concerning intuitionism raised by Chihara. In a way, Heyting, who views the existence assertions of classical mathematics as metaphysical and vague, can be interpreted as putting forward a program which simply banishes such assertions from mathematics. Heyting constructs a logical system in which the existential quantifier is accounted for in terms of assertability conditions ('$\exists x\, P(x)$' can be asserted only if one can specify a mathematical object o of which it can be truly said that $P(o)$). Chihara suggests that 'Intuitionism is not so much a philosophy of mathematics as it is the construction of a new kind of mathematics' (Chihara 1990, 22). He says:

> …philosophy of mathematics is, and should be, directed at understanding actual mathematics, and dealing with the philosophical problems arising out of the enormous role (actual) mathematics plays in everyday life and in science. When Heyting turned his back on classical mathematics and gave up trying to understand classical mathematical assertions, he in effect gave up philosophy of mathematics! (Chihara 1990, 23)

In a similar manner, if one simply disregards set theory and thinks of Mereology as the correct theory of classes, it seems that one no longer is concerned with foundations of what nowadays is classical mathematic, but rather one intends to develop a wholly different theory whose foundational role is quite unclear.[52]

---

[52] Peter Simons points out: "Leśniewski's alternative to set theory is based on the (justified) assumption: If you follow me in what you mean by 'member' and 'set' you will not get into logical trouble. That is true, but the proponents of inconsistent set theory were doing more than trying to stay out of trouble (which is why they got into it in the first place): they were trying to provide a logical foundation for mathematics, some of them with a view to showing at least part of mathematics to be just logic. Leśniewski's weaker system may, or, more likely, may not be adequate to that purpose." (Simons 1993, 7)

This does not mean that Mereology as a separate theory is useless. It turned out to be interesting in quite a few different fields: linguists sometimes use it when they discuss semantics of mass terms and philosophers speak of parts when they speak about persistence, perdurance and other issues discussed in modern metaphysics.

5.4 Further Readings

Papers concerned with Mereology divide into a few groups. Some of them constitute either an introduction or a philosophical discussion of the system. Others contain mostly results obtained within the axiomatic system of Mereology. Some are concerned with properties of Mereology as a logical system—among those a special role is played by papers whose authors discuss the novelty of Mereology (its critics claim that the system is very similar to other known systems and its advocates emphasize the differences). Another group of papers is concerned with systems based on Mereology. I will briefly discuss the relevant papers in this order.

Perhaps the most accessible introduction to Mereology is Sobociński (1954a). Asenjo (1977) discusses the historical role that Leśniewski's Mereology played in the development of non–classical set theories. Lorenz (1977) elaborates on the interplay between the partition of a whole into parts and the attribution of properties to an object. Probably the most up–to–date comprehensive book (in German) discussing mereologies is Ridder (2002). Another book whose part may serve as a textbook (in English) is Simons (1987), although Simons is focusing on developing his own, non-extensional mereology.

Just like it was the case with Leśniewski's other systems, some researchers devoted their attention to obtaining simpler axiomatizations of Mereology. Various formulations were offered by Lejewski (1954a, 1955, 1962). Clay (1961) is concerned with obtaining new theorems of Mereology and simplifying its axiomatic basis. Clay (1970) shows that one of axiomatizations given by Lejewski contains a redundant axiom.

Clay (1973) discusses a definition of the class operator simpler than that given initially by Leśniewski. Leśniewski himself gave the following simplification:

$$\forall a, b \, [a \varepsilon cl(b) \equiv a \varepsilon a \wedge \forall c \, (b \subseteq el(c) \equiv a \varepsilon el(c))]$$

However, his proof of its equivalence to the original definition, as originally given (Leśniewski 1930a), depends on the definition of a set (recall that definitions in Leśniewski's systems are creative). Sobociński is reported to have asked the question: does the equivalence proof depend on the definition essentially? Or, in other words: is the use of this definition in the proof creative? Clay (1973) proves that the answer to this question is negative.

The consistency of Mereology has been proven by Clay (1968). Clay constructs a model of Mereology in which the domain is taken to be the set of all real numbers whose decimal expansions contain only zeros and ones with the exception of the number 0. Name variables range over subsets of this set. The epsilon has the standard interpretation and $prt(a)$ is non–empty only if a is assigned a singleton. In this case it denotes the set of all those numbers x in the domain which for every position where the object belonging to the singleton assigned to a has a 1 in its decimal expansion, x has 1 at the same position too. In a way, this is a proof that if Ontology enriched with the axioms for real numbers is consistent, so is Mereology.

Lejewski (1969) noticed that this might be interpreted as proving consistency of a fairly simple theory which depends on the consistency of a less intuitively obvious theory. Hence, he focuses on proving the consistency of Mereology relative to a weaker subsystem of Mereology itself. The consistency of Ontology has been proven by a member of Leśniewski's seminar before the war.[53] If we interpret ontological variables as prototothetical variables,[54] and we interpret ε as the functor of conjunction, the theory becomes a sub–theory of Prototothetic.[55] Lejewski proved that if we additionally interpret el (element of) as the prototothetical assertion operator, Mereology becomes a sub–theory of Prototothetic as well.

Clay (1974b) provides (indirectly) another consistency proof. He shows how the standard Euclidean 3–dimensional topological spaces provide a model for atomless Mereology. Since the classical Mereology is its subsystem, this proves consistency of Mereology relative to topology.

Grzegorczyk (1955) argued that the models of Mereology are just complete Boolean algebras with zeros deleted.[56] In a similar manner, Kubiński in his remark published in 1968 points out that $cl(a)$ may be taken in the algebraic interpretation (i.e. when we take the language of Ontology, interpret it in Boolean algebras by

---

[53] It is not clear who gave the proof, though.

[54] That is, if we interpret name variables as propositional variables and variables of category $\frac{\sigma}{\sigma_1,\ldots,\sigma_n}$ as variables of category $\frac{\sigma'}{\sigma_1',\ldots,\sigma_n'}$ where σ' and the categories σ_i' result from σ and categories σ_i by replacing all n's by s's.

[55] Directives of Ontology become directives of Prototothetic under that interpretation as well.

[56] A Boolean algebra is a set A with two binary operators $\wedge$ (meet) and $\vee$ (join), a unary operation $\neg$ (complement) which contains two elements 0 and 1, such that for any $x, y, z \in A$ the following hold:

$$x \vee (y \vee z) = (x \vee y) \vee z \quad x \wedge (y \wedge z) = (x \wedge y) \wedge z$$
$$x \vee y = y \vee x \quad x \wedge y = y \wedge xr$$
$$x \vee (x \wedge y) = x \quad x \wedge (x \vee y) = x$$
$$x \vee (y \wedge z) = (x \vee y) \wedge (x \vee z) \quad x \wedge (y \vee z) = (x \wedge y) \vee (x \wedge z)$$
$$x \vee \neg x = 1 \quad x \wedge \neg x = 0$$

The ordering relation is put on a Boolean algebra by taking:

$$a \leq b \Leftrightarrow a = a \wedge b$$

Now, we say that a subset B of an Boolean algebra A has a supremum if and only if there is an $x \in A$ such that for all $y \in B$ $y \leq x$, and that for any $z \in B$ for which $y \leq z$ for any $y \in B$, it is the case that $x \leq z$. We say that A is a complete Boolean algebra if and only if every subset of A has a supremum (in A).

assigning subsets of the algebra to name variables) to be the least upper bound of the a's. These claims were criticized by Clay (1974a) on the point that what Grzegorczyk (or Kubiński, for that matter) reconstructed was not Leśniewski's Mereology. First of all, variables in the system discussed by Grzegorczyk were individual variables. Secondly, Grzegorczyk's quantification was first–order only. Another difference is that mereological models can be empty, whereas a complete Boolean algebra has to contain non-zero elements.[57] Asenjo (1977) remarks that Grzegorczyk's analogy breaks down if one does not require that Mereology has to be atomistic (Leśniewski's Mereology did not require the existence of atoms).

Słupecki (1958) can be interpreted as a similar attack against the novelty of Mereology. Słupecki hints that Mereology does not differ formally from a certain elementary theory of partial ordering. However, he formulates it with Chwistek's simple type theory as underlying logic, which as Clay (1974a) points out, 'fails to include Ontology or any other theory of the distributive class to use as counterpoint for the notion of collective class.'

At least two straightforward modifications of Mereology are available. First, it is possible to add an axiom stating that every object is either an atom or is constructed from those atoms which are its parts, or one can add an axiom stating that no atoms exist. The former extension of Mereology has been dubbed 'Atomistic Mereology' and studied by Sobociński (1971). Both atomistic and atomless Mereologies have been provided with fairly simple axiomatizations by Clay (1975).

A system of mereology inspired by Leśniewski's Mereology has been constructed by Goodman and Leonard (1940). It is weaker than Mereology.

Richard Milton Martin attempted to develop a systematic approach to linguistic and multiple reference in the framework of a system of mereology (Martin 1988, 1992).

Some papers discuss the possible role that Mereology can play in foundations of mathematics. Clay (1965) introduces the notion of weak discreteness:[58]

$$\forall a \, [\mathbf{w\text{-}dscr}(a) \Leftrightarrow \forall b, c, d, e \, [(d\varepsilon cl(b) \wedge e\varepsilon cl(c) \wedge b \subseteq a \wedge c \subseteq a \wedge d = e) \rightarrow$$
$$\rightarrow \forall a \, (a\varepsilon b \equiv a\varepsilon c)]]$$

---

[57] By a non–zero element I mean an element which is **not only** a zero in an algebra. According to this nomenclature, an object which is both 0 and 1 is a non-zero element.

[58] It is contrasted with the strong notion of discreteness, given by

$$\forall a \, [\mathbf{dscr}(a) \equiv \forall c, d \, (c\varepsilon a \wedge d\varepsilon a \rightarrow c = d \vee c\varepsilon ex(d))]$$

where ex is defined by:

$$\forall c, d \, (c\varepsilon ex(d) \equiv \forall b \, (b\varepsilon el(c) \rightarrow \neg b\varepsilon el(d)))$$

This condition says that a possible name a is discrete iff for any two possible names that uniquely name individuals named by a, they either name the same individual, or they name objects which have no common part.

The condition states that a possible name a is weakly discrete iff for any two names that name some of the objects named by a, if they differ on the objects that they name, their (mereological) classes are different. He shows that if the condition of weak discreteness is satisfied, classes behave in certain aspects like distributive sets.[59] Clay proves an analogue of (5.35) (see p. 20), but instead of using the strong discreteness in the antecedent, he requires only that a name be weakly discrete. He also proved that if two equinumerous names a and b are weakly discrete, there are as many sets of a's as there are sets of b's (see Definition II on p. 9 for details regarding Leśniewski's understanding of sets). The crucial result proven by Clay, however, is the meta–theorem saying that for any formula ϕ involving only $a_1, \ldots, a_n$ as name variables, equinumerosity and 'denoting fewer objects' operator, and finally, the *set* operator, any conditional with σ as its consequent and a formula which states strong discreteness of $a_1, \ldots, a_n$:

$$\forall a_1, \ldots, a_n \, (\mathbf{dscr}(a_1) \wedge \cdots \wedge \mathbf{dscr}(a_n) \rightarrow \phi)$$

is equivalent to a conditional with the same consequent, but whose antecedent states the weak discreteness of $a_1, \ldots, a_n$:

$$\forall a_1, \ldots, a_n \, (\mathbf{w\text{-}dscr}(a_1) \wedge \cdots \wedge \mathbf{w\text{-}dscr}(a_n) \rightarrow \phi)$$

Leśniewskian scholars have been concerned with related issues for a while. Sobociński is reported (Clay 1972) to have proven that if a is discrete (in the strong sense), then if a names finitely many objects, there are finitely many sets of a's. Clay himself (1972) proves that this condition holds even without the assumption of discreteness of a.

Słupecki (1958) attempted to modify Mereology in order to obtain a theory that could play the role of standard set theory but which would not commit one to sets. As it turns out, this strategy is not devoid of difficulties. Sect. 7.8 contains a wider discussion of this subject.

References

Asenjo, F. (1977). Leśniewski's work and nonclassical set theories. *Studia Logica, 34*(4), 249–255.
Cantor, G. (1877). Mitteilungen zur Lehre vom Transfiniten. *Zeitschrift für Philosophie und Philosophische Kritik, 91*, 81–125.
Chihara, C. S. (1990). *Constructibility and mathematical existence*. Oxford: Oxford University Press.
Clay, R. (1961). Contributions to mereology. PhD thesis, University of Notre Dame.
Clay, R. (1965). The relation of weakly discrete to set and equinumerosity in Mereology. *Notre Dame Journal of Formal Logic, 6*, 325–340.

---

[59] A problem with this approach is that if one thinks about classes mereologically, a fairly elaborate story is required to explain why exactly the condition should be accepted. Clay does not provide such an explanation.

Clay, R. (1968). The consistency of Leśniewski's mereology relative to the real numbers. *Journal of Symbolic Logic*, *33*, 251–257.

Clay, R. (1970). The dependence of a mereological axiom. *Notre Dame Journal of Formal Logic*, *11*(4), 471–472.

Clay, R. (1972). On the inductive finiteness in mereology. *Notre Dame Journal of Formal Logic*, *13*, 88–90.

Clay, R. (1973). Two results in Leśniewski's mereology. *Notre Dame Journal of Formal Logic*, *14*, 559–564.

Clay, R. (1974a). Relation of Leśniewski's mereology to boolean algebra. *The Journal of Symbolic Logic*, *39*(4), 638–648.

Clay, R. (1974b). Some mereological models. *Notre Dame Journal of Formal Logic*, *15*, 141–146.

Clay, R. (1975). Single axioms for atomistic and atomless mereology. *Notre Dame Journal of Formal Logic*, *16*(3), 345–351.

Fraenkel, A. (1928). *Einleitung in die Mengenlehre*. Berlin: Springer.

Fraenkel, A., Bar-Hillel, Y., & Levy, A. (1973). *Foundations of set theory*. Amsterdam: North Holland.

Goodman, N. (1966). *The structure of appearance*. Cambridge: Harvard University Press.

Goodman, N., & Leonard, H. (1940). The calculus of individuals and its uses. *The Journal of Symbolic Logic*, *5*, 45–55.

Grzegorczyk, A. (1955). The systems of Leśniewski in relation to contemporary logical research. *Studia Logica*, *3*, 77–95.

Kearns, J. (1962). Lesniewski, language, and logic. PhD thesis, New Haven: Yale University.

Kotarbiński, T. (1929). *Elementy Teorii Poznania, Logiki Formalnej i Metodologii Nauk [Elements of the theory of knowledge, formal logic and methodology of the sciences]*. Lwów: Ossolineum.

Kubiński, T. (1968). Uwagi o modelach systemu mereologii Leśniewskiego [Remarks about models of Leśniewski's Mereology]. *Ruch Filozoficzny*, *26*, 336–338.

Lejewski, C. (1954a). A contribution to Leśniewski's mereology. *Polish Society of Arts and Sciences Abroad*, *5*, 43–50.

Lejewski, C. (1955). A new axiom for mereology. *Polish Society of Arts and Sciences Abroad*, *6*, 65–70.

Lejewski, C. (1962). A note on a problem concerning the axiomatic foundations of mereology. *Notre Dame Journal of Formal Logic*, *4*, 135–139.

Lejewski, C. (1969). Consistency of Leśniewski's mereology. *The Journal of Symbolic Logic*, *34*(3), 321–328.

Leśniewski, S. (1914a). Czy klasa klas, niepodporządkowanych sobie, jest podporządkowana sobie? *Przegląd Filozoficzny*, *17*, 63–75. [Is a class of classes not subordinated to themselves, subordinated to itself?, (Leśniewski,1991, 115–128)].

Leśniewski, S. (1916). Podstawy ogólnej teoryi mnogości I. Prace Polskiego Koła Naukowego w Moskwie, 2. [Foundations of the general theory of sets I, (Leśniewski,1991, 129–173)].

Leśniewski, S. (1927). O Podstawach Matematyki, Wstęp. Rozdział I: O pewnych kwestjach, dotyczących sensu tez 'logistycznych'. Rozdział II: O 'antynomji' p. Russella, dotyczącej 'klasy klas, nie będących własnemi elementami'. Rozdział III: O różnych sposobach rozumienia wyrazów 'klasa' i 'zbiór'. *Przegląd Filozoficzny*, *30*, 164–206. [On the foundations of mathematics. Introduction. Ch. I. On some questions regarding the sense of the 'logistic' theses. Ch. II. On Russel's 'antinomy' concerning 'the class of classes which are not elements of themselves'. Ch. III. On various ways of understanding the expression 'class' and 'collection' (Leśniewski,1991,174–226)].

Leśniewski, S. (1928). O podstawach matematyki, Rozdział IV: O podstawach ogólnej teoryj mnogości I. *Przegląd Filozoficzny*, *31*, 261–291. [On the foundations of mathematics. Ch. IV On 'Foundations if the general theory of sets. I', Leśniewski, 1991, 227–263].

Leśniewski, S. (1929b). O podstawach matematyki, Rozdział V: Dalsze twierdzenia i definicje 'ogólnej teorji mnogości' pochodzące z okresu do r. 1920 włącznie. *Przegląd Filozoficzny*, *32*, 60–101. [On the foundations of mathematics. Ch. V. Further theorems and definitions of the 'general theory of sets' from the period up to the year 1920 inclusive, (Leśniewski,1991, 264–314)].

Leśniewski, S. (1930a). O podstawach matematyki, Rozdział VI: Aksjomatyka 'ogólnej teorji mnogości', pochodząca z r. 1918. Rozdział VII: Aksjomatyka 'ogólnej teorji mnogości', pochodząca z r. 1920. Rozdział VIII: O pewnych ustalonych przez pp. Kuratowskiego i Tarskiego warunkach, wystarczających i koniecznych do tego, by p było klasą p-tów a. Rozdział IX: Dalsze twierdzenia 'ogólnej teorji mnogości', pochodzące z lat 1921–1923. *Przegląd Filozoficzny*, *33*, 77–105. [On the foundations of mathematics. Ch. VI. The axiomatization of the 'general theory of sets' fro the year 1918. Ch. VII. The axiomatization of the 'general theory of sets' from the year 1920. Ch. VIII. On certain conditions established by Kuratowski and Tarski which are sufficient and necessary for *P* to be the class of objects *A*. Ch. IX. Further theorems of the 'general theory of sets' from the years 1921–1923, (Leśniewski 1991, 315–349)].

Leśniewski, S. (1931a). O podstawach matematyki, Rozdział X: Aksjomatyka 'ogólnej teorji mnogości pochodząca z r. 1921. Rozdział XI: O zdaniach 'jednostkowych' typu '*Aε b*'. *Przegląd Filozoficzny*, *34*, 142–170. [On the foundations of mathematics. Ch. X. The axiomatization of the 'general theory of sets' from the year 1921. Ch. XI. On 'singular' propositions of the tyle '*Aε b*', (Leśniewski,1991, 350–382)].

Lorenz, K. (1977). On the relation between the partition of a whole into parts and the attribution of properties to an object. *Studia Logica*, *36*, 351–362.

Luschei, E. (1962). *The logical systems of Leśniewski*. Amsterdam: North-Holland.

Martin, R. M. (1988). *Metaphysical foundations: Mereology and metalogic*. Analytica. Munich: Philosophia Verlag.

Martin, R. M. (1992). Logical semiotics and mereology. In A. Eschbach (Ed.), *Foundations of Semiotics*, (Vol. 16). Amsterdam: John Benjamins Publishing Company.

Ridder, L. (2002). *Mereologie. Ein Beitrag zur Ontologie und Erkenntnistheorie*. V.Klostermann

Simons, P. (1987). *A study in ontology*. Oxford: Oxford University Press.

Simons, P. (1993). Nominalism in Poland. In F. Coniglione, R. Poli & J. Woleński (Eds.), *Polish scientific philosophy: The Lvov-Warsaw school* (pp. 207–231). Amsterdam: Rodopi.

Simons, P. (1995). Lesniewski and ontological commitment. In D. Miéville & D. Vernant (Eds.), *Stanislaw Lesniewski Aujourd'hui, number 16 in Recherches Philosophie, Langages et Cognition*, (pp. 103–119). Université de Grenoble.

Słupecki, J. (1958). Towards a generalized mereology of Leśniewski. *Studia Logica*, *8*, 131–154.

Sobociński, B. (1949b). L'analyse de l'antinomie Russellienne par Leśniewski. *Methodos*, *1–2*(1, 2, 3; 6–7), 94–107, 220–228, 308–316; 237–257. [translated as "Leśniewski's analysis of Russell's paradox" (Srzednicki and Ricky, 1984, 11–44)].

Sobociński, B. (1954a). Studies in Leśniewski's Mereology. *Yearbook for 1954–55 of the Polish Society of Arts and Sciences Abroad*, *5*, 34–48.

Sobociński, B. (1971). Atomistic mereology. *Notre Dame Journal of Formal Logic*, *12*, 89–103.

Whitehead, A. and Russell, B. (1910–1913). *Principia Mathematica*. Cambridge: Cambridge University Press. [Reprinted in 1960].

Chapter 6
Leśniewski and Definitions

Chapter Written with Severi K. Hämäri

Abstract We indicate, document and attack a popular misunderstanding about Leśniewski and definitions, according to which it was Leśniewski who came up with the consistency and conservativeness requirements on definitions. After some preliminaries we explain the origins of the folklore, elaborate on Leśniewski's unusual style which contributed to the obscurity of his works, explain why Leśniewski's definitions are creative, and what Leśniewski's rules for definitions actually accomplished. Finally, we argue that most of the credit on the Polish ground should go to Łukasiewicz and Ajdukiewicz instead.

6.1 Introductory Remarks

A theory of definitions which places the eliminability and conservativeness requirements on definitions is usually (following Belnap 1993) called the *standard theory*. In this chapter[1] we look at a persistent myth which credits this theory to Leśniewski.[2] After a brief survey of its origins, we show that the myth is highly dubious:

- No place in Leśniewski's published or unpublished work is known where the standard conditions are discussed.
- Leśniewski's own logical theories allow for creative definitions.
- Leśniewski's celebrated 'rules of definition' lay merely syntactical restrictions on the form of definitions: they do not provide definitions with such meta-theoretical requirements as eliminability or conservativeness.

---

[1] This chapter is based on a paper (Urbaniak and Hämäri 2012), which contains more details and a collection of translations of main source texts.

[2] This common belief has lingered at least since the publication of Suppes (1957) (but only Nemesszeghy and Nemesszeghy (1977) try to bring up some evidence to corroborate it).

R. Urbaniak, *Leśniewski's Systems of Logic and Foundations of Mathematics*,
Trends in Logic 37, DOI: 10.1007/978-3-319-00482-2_6,
© Springer International Publishing Switzerland 2014

On the positive side, we point out that among the Polish logicians, in the 1920s and 1930s, a study of these meta-theoretical conditions is more readily found in the works of J. Łukasiewicz and K. Ajdukiewicz.

This issue bears on how we are to estimate Leśniewski's impact on Tarski. Betti (2008a) argues that the forerunner of Tarski's semantical investigations was not Leśniewski, but rather Ajdukiewicz. Although Betti does not discuss the history of the desiderata put on definitions, Hodges (2008) asks how Leśniewski's theories affected Tarski's views on definitions. He concludes that '[i]t's impossible to measure how far Leśniewski's other views[3] on definitions influenced Tarski without establishing what those other views were, and this is difficult' (103).

Below (in Sects. 6.6 and 6.7) we try to shed more light on what 'those other views were'. Even though many of these issues are well known among Leśniewski scholars,[4] the persistence of the above-mentioned folklore indicates that the wider audience might not be as familiar with them. This unfamiliarity is not completely surprising, given Leśniewski's unapproachable style (see Sect. 6.4) and, until recently, the lack of comprehensive and tangible secondary sources on his theory of definitions.[5] Two separate issues are at play here: on one hand, we may ask who endorsed the theory; on the other, who analyzed the conditions. We argue (in Sect. 6.7) that Leśniewski's 'rules of definitions' ensured eliminability but *allowed for creativity* without him commenting on why he proceeded this way. Instead, we establish (in Sect. 6.8) that Łukasiewicz employed the conservativeness requirement in an argument against Leśniewski, and that Ajdukiewicz managed to develop a rather elaborate account of those requirements that forms the grounds of the standard theory. Finally, we dismantle the evidence which Nemesszeghy and Nemesszeghy (1977) use to argue that the theory is Leśniewski's. We hope that our treatment here paints an accessible picture of Leśniewski's position on definitions for those not yet acquainted with it.[6]

---

[3] Hodges argues that there is a link between Leśniewski's requirement of 'irresistible intuitive validity' and Tarski's 'wymóg trafności' (i.e. the material adequacy requirement) [102–103, 114].

[4] See for instance (Rickey 1975a; Simons 2008b).

[5] For instance, although Luschei (1962, 36) testifies that 'Leśniewski's rules for definition... are among his most important scientific contributions, and need to be rescued from comparative oblivion', and further explains that he knows of 'no other rules comparable in adequacy and rigor of formalization to Leśniewski's directives of definition, more comprehensive and exact even than Frege's' [*ibidem*], he leaves the nature of those rules rather unexplained. He brings them up only in his rendering of Terminological Explanations, especially in T.E. XLIV. This terminological explanation starts with 'A is legitimate as propositive definition immediately after thesis B of this system if and only if the following eighteen conditions are fulfilled...' and continues with a rather literal representation of what Leśniewski himself said about the conditions.

[6] Since we are here concerned only with the Poles of Lvov-Warsaw school, we can only hint at the discussions of e.g. Kant's, Mill's, Frege's, Peano's, and Russell's roles in the early developments of the standard account. (We have a few remarks in Sects. 6.3 and 6.5, though.) Decent introductions to theories of definitions are Gupta (2009) and Abelson (1967) and the monographs (Dubislav 1981; Robinson 1954). Most original texts prior to Frege are in Sager (2000) (although it lacks the relevant passages from *Port-Royal Logic*. Beck (1956) seems still the authority on Kant's treatment of definitions. Shieh (2008) has extensive Frege references (both primary and secondary) as well

6.2 Basic Notions

The standard theory, in contrast to theories about teachers', lawyers', legislators', or lexicographers' definitions, is concerned with definitions in the logicians' sense: a definition is a formula (or a set of such) which, relative to a formal theory and a language, fixes the meaning of new symbols occurring in it. (We can ignore the otherwise important distinction between explicit and implicit definitions here. Also, we are not considering meta-theoretical abbreviations, but object level "acronymic" definitions which introduce new symbols into the object language).

As Belnap (1993, 119) reasons, 'a definition of a word should explain *all* the meaning that [the] word has, and... it should do *only* this and nothing more.' He suggests that a natural connection with these intuitive desiderata might be the reason why the standard theory placed eliminability and conservativeness as the criteria[7] of acceptable definitions. He also argues that the criterion of eliminability ascertains that the intuition about 'all the meaning' is satisfied while the criterion of conservativeness, i.e. non-creativity, ensures the satisfaction of the 'only' part.

Let us look into some technical details. In deductive terms eliminability is defined as follows. Let T be a theory, L its language and Δ a set of formulas in an extended language $L' \supseteq L$. Then, symbols in $L' \setminus L$ are *eliminable* (i.e. "Δ is eliminable") *in $T \cup \Delta$ with respect to L* if and only if for any formula ϕ in language L' there exists a formula ψ in L such that $T \cup \Delta \vdash \phi \equiv \psi$. In other words, every expression in the extended language has to be "translatable" into the old language. Conservativeness, on the other hand, is defined as follows. Let T, Δ, L and L' be as above. Δ is *conservative over T* if and only if for all formulas ϕ in the language L, if $T \cup \Delta \vdash \phi$ then $T \vdash \phi$. This means that no new claim that can be formulated in the original language becomes provable once the definition is introduced. As we can see, according to the standard view a set of formulas Δ is a definition only with respect to some theory and some language; there are no general syntactical criteria of definitions here.

(Footnote 6 continued)

as a good introduction to his theory of definitions. On Peano's groups' treatment of definitions and definability in mathematics there appears to be only a few sources in English. Grattan-Guinness (2000) is a notable exception. His book is a valuable source on the other players as well, just search his index for 'definition'. Anyone interested in *the Peanists* should look at Padoa (1900), which is one of the most important texts in early model theory and the problem of definability. Before E. Beth's breakthrough in the 1950s, Tarski was one of the few working on Padoa's method, see Tarski (1934) and Hodges (2008). Kennedy (1973) sketches the impact Peano's theory of definition had on young Russell. Dubislav's role in the development of the standard theory might turn out to be important. This German positivist and logician published two editions of his book *Die Definition* (Dubislav 1981) in the late 1920s and the third edition in 1931. He discusses creative (*schöpferischen*) definitions in considerable detail. We know that at least Koj (1987) talks about Dubislav's impact on Ajdukiewicz. Grattan-Guinness (2000, 486, 519–520) provides a description of Dubislav's work in general. For anyone interested, most present-day issues are examined in the collection (Fetzer et al. 1991).

[7] We do not use the word 'criterion' in a technical "Wittgensteinian" sense, but as a synonym of 'condition' and 'requirement'.

To avoid a problem arising from the possibility of two separately conservative (or eliminable) but jointly creative (non-eliminable) definitions, the definitions have to be introduced in stages. Each subsequent stage has to be eliminable and conservative over the previous ones. We cannot go into more details of the standard theory. For the particularities see Suppes (1957, Chap. 8), Mates (1972, 197–203) and especially Belnap (1993).[8]

As it turns out, eliminability (even supposing consistency) is not a sufficient criterion for a successful definition. Definitional extensions might differ on a fundamental level from the original theory, e.g. they might lack a model with cardinality in the spectrum of the original theory.

Without the criterion of conservativeness we would also run the risk of Prior's (1960) skepticism or 'tonktitis', as Belnap (1962) called it. Prior argues that the rules of deduction cannot function as definitions that explain all the meanings of the logical constants: there are inconsistent sets of rules (that is, rules which lead to triviality). But there is more here than meets the eye. Belnap replies that there is a difference between rules that can function as definitions and rules that cannot: the rules that define meaning are conservative over the underlying logic. Therefore, the criterion of conservativeness is important to the proof-theoretic meaning theory as well (see also Hacking 1979).

As we will explain in Sect. 6.6, Leśniewski's definitions (e.g. in his theory called *Ontology*) satisfy the eliminability but not, in general, the conservativeness requirement (yet we have pretty good reasons to think the system is consistent (Słupecki 1955)). Why, then, was the standard theory ascribed to him?

6.3 On the Origins of the Folklore

As far as we can tell, no publication by Leśniewski is the origin of the folklore, and most book passages and articles on definitions between the 1930s and the 1960s contain nothing about Leśniewski. For instance, even in (Łukasiewicz 1963, 31–33) where Leśniewski's views on definitions are implicitly criticized, none of these views are explicitly attributed to him. (We will come back to Łukasiewicz's lecture notes, originally presented in 1928–1929, in Sect. 6.8.)

A reference to Leśniewski's position, however, can be found at least in Łukasiewicz (1928b), Tarski (1941), Mostowski (1948), Kelley (1955), Church (1956), Suppes (1957) and Ajdukiewicz (1936) (in an introduction written in 1960). We will show in Sect. 6.8 that Lukasiewicz and Ajdukiewicz have not contributed to the folklore. Let us look a bit closer at the other sources.[9]

---

[8] Došen and Schroeder-Heister (1985) discuss slightly different notions (which are used in Belnap (1962): conservativeness and uniqueness. They show that this pair of conditions has interesting properties (e.g. they are dual to each other in certain contexts).

[9] For the sources on Leśniewski's views, see also (Hodges 2008, 103–5).

In Tarski's *Introduction to Logic* Tarski (1941) we find that:

> ...the present day methodology endeavors to replace subjective scrutiny of definitions... by criteria of an objective nature, in such a way that the decision regarding the correctness of given definitions... would depend *exclusively upon their structure, that is, upon their exterior form.* For this purpose, special rules of definition... are introduced. [These rules] tell us what *form the sentences should have* which are used as definitions... *each definition has to be constructed in accordance with the rules of definition*... [In footnote:] one of [Leśniewski's] achievements is an exact and exhaustive formulation of the rules of definition. (Tarski 1941, 123, n. 4)[10]

Although Tarski does not explicate what Leśniewski's 'rules of definition' are, we agree with Hodges (2008) that it is clear from the context that Tarski means some syntactical criteria for good definitions. There is no mention of meta-theoretical requirements in the passage. As we will see later, Tarski's remark is adequate. Still, Tarski cites no works and states no explicit rules by Leśniewski and this might have caused some confusion.

Church when writing about the object language definitions in *Introduction to Mathematical Logic* Church (1956) states that he...

> ...agrees with Leśniewski that, if such definitions are allowed [in the object language], it must be on the basis of *rules of definition*, included as a part of the primitive basis of the language and as precisely formulated as we have required in the case of the formation and transformation rules... Unfortunately, authors who use definitions in this sense have not always stated rules of definition with sufficient care... On the other hand, once the rules of definition have been precisely formulated, they become at least theoretically superfluous, because it would always be possible to oversee in advance everything that could be introduced by definition, and to provide for it instead by primitive notations included in the primitive basis of the language... Because of the theoretical dispensability of definitions in [this] sense... we prefer not to use them... (Church 1956, 76, n. 168).

Here it is clear that the mentioned rules of definition are taken to be syntactical and that object language definitions should at least be eliminable. Although it appears that Church is silent about creativity, it is nonetheless plausible that the definitions which he would allow are non-creative, for he states that the correct rules for definitional introduction of symbols would make the system interchangeable with another one that lacks them. Church could well have adhered to the standard theory.

What, then, is Church attributing to Leśniewski? An interpretation which reads Leśniewski as embracing the 'superfluity' of object language definitions is of little credibility. Thus we are left with another more plausible but rather vague interpretation on which Leśniewski wanted definitions to be governed by precise rules. On

---

[10] The emphases are changed. There are some small points pertinent to the history of this passage in Tarski's *Introduction*. The German edition (Tarski 1937), which otherwise seems to agree with the translation, does not credit Leśniewski with the formulation of the rules of definition. Hodges (2008) notes that the attribution of 'an exact and exhaustive formulation of the rules of definition' appeared in the English edition of 1941. He suggest that the change might be 'a mark of respect for a teacher who had died just 2 years earlier.' [103]. Also, in the fourth edition of 1994 'an exact' is replaced by slightly less emphatic 'a precise'. It is not clear whether it was Tarski's decision, Jan Tarski's improvement of his father's style, or John Corcoran's choice of a more appropriate word.

this reading Church (1956) (in a way, just like Tarski) contributes to the folklore not by what he says, but rather by what he leaves unexplained.

The same is true about Mostowski. The most explicit remark about Leśniewski to be found there is:

> The need of a precise formulation of the rule of definitions has been strongly emphasized by Leśniewski. He gave a precise formulation of this rule with respect to the systems he constructed. (Mostowski 1948, 251)

Kelley (1955) and Suppes (1957), on the other hand, subscribe to the folklore. Kelley states in an appendix that he implicitly posits 'an axiom scheme for definition' and that...

> ...the axiom scheme of definition is in the fortunate position of being justifiable in the sense that, if the definitions conform with the prescribed rules, then no new contradictions and no real enrichment of the theory results. These results are due to S. Leśniewski. (Kelley 1955, 251, n.)

Here Kelley is slightly vague: what would he count as a 'real enrichment'? If, as is natural to assume, he means inferential creativity, then he might be the first to state the folklore. (He is partly right since Leśniewski's rules would ascertain the consistency of the introduced definitions within a consistent theory, even though the relative consistency proof is not due to Leśniewski.) Kelly presents no source for his claim.

Suppes (1957) seems far more explicit than Kelley. He states that:

> it is not intended that a definition shall strengthen the theory in any substantive way. The point of introducing a new symbol is to facilitate... investigation... but not to add to... [the] structure [of a theory]. Two criteria which make more specific these intuitive ideas about the character of definitions are that (i) a defined symbol should always be eliminable from any formula... and (ii) a new definition does not permit the proof of relationships among the old symbols which were previously unprovable; that is, it does not function as a creative axiom. [In a footnote:] These two criteria were first formulated by the Polish logician S. Leśniewski... he was also the first person to give rules of definition satisfying the criteria. (153)

As we can see, Suppes states the standard theory; and there is no doubt about what he credits to Leśniewski. Suppes claims that Leśniewski (a) had 'formulated' the conditions and (b) had laid these requirement on some 'rules of definitions'.

The former claim is dubious, bearing in mind Frege's remarks on non-creativity and eliminability of definitions in 1914:

> Now when a simple sign is thus introduced to replace a group of signs, such a stipulation is a definition. The simple sign thereby acquires a sense which is the same as that of the group of signs. *Definitions are not absolutely essential to a system. We could make do with the original group of signs.* The introduction of a simple sign adds nothing to the content; it only makes for ease and simplicity of expression.
>
> ...
>
> A sign has a meaning once one has been bestowed upon it by definition, and the definition goes over into a sentence asserting an identity. Of course the sentence is really only a tautology and does not add to our knowledge. It contains a truth which is so self-evident

that it appears devoid of content, and yet in setting up a system it is apparently used as a premise. I say apparently, for what is thus presented in the form of conclusion makes no addition to our knowledge; all it does in fact is to effect an alteration of expression, and we might dispense with this if the resultant simplification of expression did not strike us as desirable. *In fact it is not possible to prove something new from a definition alone that would be unprovable without it. When something that looks like a definition really makes it possible to prove something which could not be proved before, then it is no mere definition but must conceal something which would have either to be proved as a theorem or accepted as an axiom.* (Frege 1914, 208)

Even earlier (he has remarks about non-creativity already in *Begriffschift* (1879, 55), see also Shieh 2008, 994), not to mention the fact that the basic ideas behind eliminability as well as non-creativity were, arguably, known at least by Blaise Pascal. The case of eliminability is clear, since the possibility of substituting mentally the defined term with the *definiens* in all contexts is mentioned many times in *De L'esprit Géométrique* (1814b, e.g. 127). But the case of non-creativity needs an argument. Pascal not only demands that the sentences containing the defined word must be translatable to ones not containing it. He also demands that all proofs containing the defined word must be translatable into proofs not containing the word [otherwise proofs would not be persuasive (Pascal 1814a, 161)]. Therefore, one might argue (just like in the case of Church) that Pascal, if asked, would have said that definitions are non-creative. But the cases are not identical, and we have here an open question: would Pascal allow for a definition to act as an axiom with existential import?[11] If the answer to this query is positive, then Pascal would have accepted creative definitions nonetheless.[12] But, without doubt, non-creativity was known to Mill (1869, 100) who lampooned the idea of using definitions as premises in any other context than when we are dealing with words.

The failure of claim (b) is slightly less obvious, since, as we will see in Sect. 6.4, Leśniewski's idiosyncratic style is difficult to follow. However, given that Leśniewski's logical systems abide by his rules of definitions and yet definitions in those systems are creative (see Sect. 6.6), (b) also turns out to be rather implausible. It is remarkable that both Hodges (2008, 104) and Rickey (1975b) review the points (a) and (b) of the folklore rather similarly. We discuss their critiques in Sect. 6.5.

Further, in Suppes (1957) there is no citation supporting claims (a) and (b) (According to Hodges (2008, 105), Suppes could not name the reference when Hodges enquired about it in 1996 and later.). As far as we know, and contrary to some claims (see Sect. 6.5), no one has found any discussion of the conditions in Leśniewski's works. It is probable that none of the texts used by Suppes[13] attributed

---

[11]   e.g. Hobbes seems to have held such a position, see (Abelson 1967, 318). Think of a definition of number zero: zero is the number which has no predecessor. Now, one strategy to eliminate 'zero' would be would be to apply existential generalization: there is a number which has no predecessor. Such an eliminative definition of zero would be creative if used as a premiss.

[12]   Compare (Abelson 1967, 319).

[13]   Tarski (1941) is among Suppes's references, whereas Kelley (1955) and Church (1956) are not.

the standard theory to Leśniewski. Most other authors cite either Suppes or Belnap (1993), whose source is also Suppes.[14]

In general, it seems that already before the mid-1950s writers of introductory books wrote about Leśniewski and his so-called rules of definition. Every time this happened, the rules were rather mentioned than explicitly described. This tendency has continued: Leśniewski's rules for definition receive no detailed discussion even in Woleński's classic book Woleński (1985) whose English translation serves the role of the standard reference when it comes to the Polish school. This unwillingness to explicate turns out to be nothing surprising given Leśniewski's "Byzantine" writing style.

6.4 On Leśniewski's Idiosyncrasies

One of the reasons why the actual nature of Leśniewski's rules is hardly known is their rather convoluted form. Therefore, the reader might find a quick tour of his approach useful. Once we briefly survey these issues, we move on to the later development of the folklore. Then, we examine the reasons why Leśniewski's systems need creative definitions and what his rules for definitions actually state.

In his papers Leśniewski uses at length a truly idiosyncratic terminology to define the rules of inference and the rules of definitions for his systems by means of what he calls *terminological explanations* (T.E.). Chronologically, the first paper where he employs terminological explanations to talk about definitions is from (1929a) and his (1931b) is the second one.[15] In the former he presents rules for his system of Protothetic and in the latter he gives rules for a simpler system of classical propositional logic (here he employs Łukasiewicz's axiomatization).

Few would call Leśniewski's style *user-friendly*. For instance, the first terminological explanation in Leśniewski (1931b), probably the most straightforward one, elaborates on the composition of an expression A from a "collection" a of symbols. The lower-case variables behave like the name variables, which will be described in detail in Sect. 6.6, and capitalized variables stand for singular terms. (Although it is quite natural to interpret Leśniewski's name variables set theoretically, as collections, it is not in accord with his original ideas, hence the scare quotes. Perhaps, a slightly

---

[14] "I learned most of the theory first from Suppes (1957), who credits Leśniewski ..." (Belnap 1993, 117). Here he refers to the same quotation in Suppes above.

[15] Publication dates are not perfectly representative of when Leśniewski came up with various things, for he often tended to keep his papers in the drawer for a while; so it seems that Mereology dates back to 1916, Ontology dates back to around 1920 (see e.g. Leśniewski 1931a, 367), and Prototothetic dates back at least to 1923; it is not clear whether the fact that the systems were constructed in those years means that full-blown ready-to-print descriptions of those systems are the same age (for more details, see Urbaniak (2008b, 71–74, 105–107, 140–142).

more plausible reading takes them to be plural variables, but we do not need to get into these details here.)[16] The original formulation of T.E. I goes as follows:

> I say of object A that it is (the) complex of (the) a if and only if the following conditions are fulfilled:
>
> (1) A is an expression;
> (2) if any object is a word that belongs to A, then it belongs to a certain a;
> (3) if any object B is a, and object C is a, and some words that belongs to B belongs to C, then B is the same object as C;
> (4) if any object is a, then it is an expression that belongs to A. (Leśniewski 1931b, 631).

The underlying intuition is that for A to be composed of expressions a, (1) A has to be an expression (2) composed of words which occur in an expression which is a only, where (3) expressions a have no words in common, and (4) contain no expression that does not occur in A.[17]

This is only the first terminological explanation in Leśniewski (1931b) and they get more complicated; elsewhere (1929a) he is even less reader-friendly: he presents his terminological explanations in his full-fledged, idiosyncratic, formalized meta-language with little explanation in natural language.[18] For example, the tenth termi-nological explanation in (1929a) looks like this:

$$\forall A\, [A \,\varepsilon\, qnr1 \equiv$$
$$\exists B\, (B \,\varepsilon\, qntf \wedge B \,\varepsilon\, ingr(A) \wedge lingr(A) \,\varepsilon\, ingr(B)) \wedge$$
$$\wedge \exists B\, (B \,\varepsilon\, sbqntf \wedge B \,\varepsilon\, ingr(A) \wedge Uingr(A) \,\varepsilon\, ingr(B)) \wedge$$
$$\wedge \forall B, C\, (B \,\varepsilon\, qntf \wedge Bingr(A) \wedge C \,\varepsilon\, sbqntf \wedge C \,\varepsilon\, ingr(A) \wedge$$
$$\wedge lingr(A) \,\varepsilon\, ingr(B) \wedge Uingr(A) \,\varepsilon\, ingr(C) \rightarrow$$
$$\rightarrow A \,\varepsilon\, Compl(B \cup C))]$$

All of this only states the necessary and sufficient conditions for an expression A to be a quantified formula (i.e. that (1) there is a quantifier, which contains the bound variables, which occurs in A, and whose first word is the first word of A; (2) there is a range of a quantifier which occurs in A such that the last word occurring in A occurs in the range of this quantifier; and (3) for any two expressions B and C such that B is a quantifier occurring in A and C is a range of a quantifier and occurs in A, if the first word in A occurs in B and the last word of A occurs in C, A is the result of the composition of B and C).

Leśniewski uses this kind of strategy at length to define the systems under investigation. In (1931b) he presents an axiom for the classical propositional calculus,

---

[16] For more details pertaining to the philosophical issues related to Leśniewski's variables and quantifiers see (Urbaniak 2008b, Chap. 7).

[17] For a slightly elaborate explanation of what Leśniewski means by 'words', he sends the reader to his 1929a paper.

[18] Leśniewski, however, does not think of it as a formal system *sensu stricto*.

formulates the rules of inference (detachment and substitution), and syntactically defines the shape of correct definitions.

For instance, Ajdukiewicz (1928, 51), who in general agrees that Leśniewski's system 'is the only system with precisely formulated rules for definitions', when comparing Leśniewski's rules for definitions with those of Frege emphasized that the main difference between their views is that Leśniewski explicitly required that the *definiens* should not contain quantifiers and that it should not contain different occurrences of one and the same variable. Nothing beyond that, even if the complicated form of the description may make its content seem more elaborate.

In general, the preceding examples should suffice as an illustration of the challenge that detailed understanding of Leśniewski's rules poses, and as an explanation of the relative unpopularity of his work. An uncharitably minded reader could say that no deep intrinsic logical complexity is involved in Leśniewski's explications; after all, if all the formulation does is provide a description of what the axiom is and what rules of inference are, this can be done in a more accessible manner; and she might conclude that Leśniewski's meticulous emphasis on precision and full formalization is an overkill which, mixed with the unfamiliarity of his language, makes the effort of reading his work seem too strenuous.

This attitude is understandable. Most logicians can lead their lives tackling more intrinsically interesting logical problems without requiring this level of precision. Considering the fact that Leśniewski was publishing in the late 1920s and early 1930s, when deeper logical problems surfaced, it is no wonder his work received little attention (For example, Jordan (1945, 44), who knew Leśniewski's works well, says that reading Leśniewski's account of definition is a 'somewhat excruciating experience' for anyone 'anxious to spare themselves the valuable thought'.). But this fact does not imply that Leśniewski's approach is worthless. For example, the development of proof theory in the 1960s and 1970s required almost the same level of syntactical precision that Leśniewski embraced. Some possible reasons why he adopted such a style and stern standards are discussed in Simons (2008b).

On the other end of the spectrum, some people when faced with Leśniewski's complicated metalanguage get an impression of hidden wisdom and a feeling that more is being said than what they can grasp. They too are somewhat responsible for the long life of the myth.

6.5 Later Developments in the Folklore

Probably the most known recent paper about the standard account is Belnap (1993), where Belnap acknowledges the folklore. Here, however, we can see some caution: he is painfully aware of the lack of an original source. As we discussed earlier, his only reference is to Suppes's *Introduction to Logic*. He writes that '[t]he standard theory of definitions seems to be due to Leśniewski, who modeled his "directives" on the work of Frege, but I cannot tell you where to find a history of its development.' Belnap continues with a guess at which texts might be relevant. He supposes that the theory might be in Leśniewski (1931b), or at least somewhere in Leśniewski's *Collected works*; as far as we know, it is in neither.

Because of Belnap's reservations we see caution, for example, in Horty (2007). When Horty discusses how the fruitfulness of definitions relates to the requirements of conservativeness and eliminability in Frege's works,[19] he notes that the 'explicit formulation' of the criteria 'is generally credited to Leśniewski' and that he knows 'of no complete history of the modern theory of definition, but some historical remarks can be found in Belnap (1993)...' [34, n. 4].

On the other hand, Gupta (2009) implies that the standard account might be by Ajdukiewicz. He states that one of the present authors, R. Urbaniak, holds this stance. However, at the time, Urbaniak had only remarked on two issues concerning this debate. First, the conditions were to the best of his knowledge not formulated by Leśniewski. Second, as long as the Polish logicians are concerned, Ajdukiewicz studied the conditions. Unlike Gupta, we do not claim unqualified Polish origin of the standard theory; and, as our recent findings suggest, the priority even among the Poles might belong to another logician, Jan Łukasiewicz.

The most exciting period in the development of the folklore is the 1970s. During that time, a discussion over the admissibility of the definition of implication to be found in *Principia Mathematica* raised also a debate over the justification of the ascription of the standard requirements to Leśniewski. This debate appears to be unknown to the later authors on the history of definitions.

For example, neither Belnap (1993), nor Gupta (2009), nor Hodges (2008) refer to these papers from the 1970s. In Nemesszeghy and Nemesszeghy (1971), which started the discussion about implication in *PM* (Nemesszeghys claim that, unexpectedly, the *PM* definition of implication is creative) we find an indubitable affirmation of the folklore.[20]

Dudman (1973), in a reply, points out that the attribution is mistaken because of the priority of Frege's writings. Rickey (1975b), on the other hand, brings forth strong arguments against the folklore. Rickey criticizes especially Nemesszeghy and Nemesszeghy (1971), but notes that the essentially same argument can be put forward against Kelley (1955), and Suppes (1957). On all his points pertaining to Leśniewski, Rickey does not just rely on his own expertise, but he expresses gratitude to B. Sobociński who '[verified] all of the comments... about Leśniewski'. This is interesting because Sobociński (who was a student of Leśniewski and, after his teacher's untimely death, one of the main contributors in the study of Leśniewski's systems alongside with Lejewski) took care of Leśniewski's *Nachlass* from 1939 until it was lost around 1944 (See Simons 2008b).

---

[19] Here Horty claims that although Frege subscribed to the standard requirements, he gave no logical analysis of these conditions.

[20] Even though Nemesszeghy and Nemesszeghy do not credit Suppes, they seem to be following him. Here is the quotation from Nemesszeghy and Nemesszeghy (1971) for comparison with the above citation from Suppes (1957): "The idea that definitions should not strengthen the theory in any significant way finds expression in the following two criteria first formulated by the Polish logician S. Leśniewski: (1) a defined symbol should be always eliminable, (2) a definition should not permit the proof of previously unprovable relationships among the old symbols."

The negative part of Rickey's argument can be reconstructed as follows: he points at (a) the priority of Galileo Galilei on the notion of eliminability of definitions[21] and at (b) the priority of Pascal and Mill on the notion of non-creativity; he also claims that (c) there is no manuscript, paper or book in Leśniewski's oeuvre where these requirements are mentioned as *the criteria*; and that (d) in his theories Leśniewski utilized creative definitions freely.

Hodges (2008, 104) attacks the folklore, as expressed in Suppes (1957), by independently formulating an argument similar to Rickey's reasoning. Hodges comments in a short passage on (a) Pascal's and Porphyry's prior discussion of eliminability, on (b) the priority of Frege regarding non-creativity, on (c) the absence of textual evidence, and on (d) the fact that Leśniewski endorsed creative definitions. Hodges also claims that 'Leśniewski probably had no general theory of definitions', i.e. that Leśniewski treated definitions in a piecemeal manner, in one deductive system at a time. Nothing in our findings contradicts this statement.

Rickey in addition lays out a summary of Leśniewski's positive achievements pertaining to definitions[22]: (1) Leśniewski showed that definitions can be used on the object language level, and that the symbol '$=_{df}$' is thus superfluous; and (2) '[s]ince definitions are in the object language Leśniewski realized—and this is a valuable contribution—that it is necessary to have rules for introducing definitions.' [176]. Rickey notes that the rules Leśniewski placed on definitions in Leśniewski (1931b) ascertained eliminability and consistency.

He concludes that when Leśniewski developed his rules '[c]reative definitions were neither defined nor discussed…. However, some of the definitions introducible according to that rule are creative (relative to the particular axiom system chosen)'. (Rickey 1975b, 176). Notwithstanding the clarity of his expression here, Rickey's critique seems to have been mostly unnoticed or forgotten: as far as we know, the only ones who reacted to this criticism were Nemesszeghy and Nemesszeghy (1977). And they argued against it.

Nemesszeghy and Nemesszeghy (1977) presented the basis of their argument already in an earlier reply to Dudman.[23] In their Nemesszeghy and Nemesszeghy (1973) they write that in the previous paper they 'did *not* attribute to Leśniewski the view that all definitions should satisfy the criteria'.[24] They admit that 'Leśniewski used definitions which satisfied [eliminability] but not [non-creativity]'. On the contrary, they continue to point out that they only meant 'that the idea that definitions should not strengthen the theory in any significant way finds expression in those two criteria of Leśniewski'. In other words, the argument of Nemesszeghy and Nemesszeghy implies the exegetical difference that we mentioned in the introduction,

---

[21] Rickey cites Galilei (2001, 28) where 'mathematical definitions' are described as 'abbreviations'.

[22] He restates these contributions in Rickey (1975a). We treat these issues in Sect. 6.7.

[23] It appears unlikely that Rickey had read Nemesszeghy and Nemesszeghy (1973) before writing Rickey (1975b).

[24] They are correct in this, though the opposite reading is natural, too: see n. 20 above.

between following the requirements and studying them; and they credit Leśniewski with the latter.[25]

It is just this difference that Nemesszeghy and Nemesszeghy (1977, 111–112) accuse Rickey of "fusing and confusing" and they reiterate that they 'did think, and still think, that one can truly hold' that the '[c]onditions… can be attributed to Leśniewski.' Nemesszeghy and Nemesszeghy (1977) base this conviction on the fact that Leśniewski 'was the first, at least in modern times, to discuss and use definitions that play a creative role' which, according to them, would not have been possible without good knowledge of the eliminability and non-creativity properties as a tool of measure: 'he had a clear idea of the distinction between "creative role" and a "mere abbreviative role" of a definition, which finds expression in [the] conditions'.

Here [unlike in Nemesszeghy and Nemesszeghy (1971) or (1973)] they provide a citation to justify their claim—page 50 in Leśniewski (1929a) (that is, p. 459 in the English translation).[26] First of all, contrary to what is claimed, the passage referred to does not support the claim (as we argue *in extenso* in Urbaniak and Hämäri 2012).

Secondly, as we argue in Sect. 6.7, the only meta-theoretical principles that Leśniewski mentions as guiding his position on definitions were consistency and the avoidance of meta-linguistic treatment of definitions; therefore it seems that he was interested in the correct syntactic form of definitions only. Finally, there is evidence, which we discuss in Sect. 6.8, that even though it was Leśniewski's work on definitions which prompted the study of the standard requirements in Poland it was Łukasiewicz who noticed the importance of the creativity of definitions, and that (as far as we know) it was Ajdukiewicz who first studied the conditions systematically (at least in Poland).

We think that the Nemesszeghys attribution of the standard requirements to Leśniewski, even only as a measure on definitions, is an overstatement. His role in the development of theory of definitions is important, but to claim on his behalf the invention of the standard conditions just because he may have been aware of them is to stretch the facts. Besides, Frege has technically at least as good a claim for them as Leśniewski, and Frege has priority. We know that Leśniewski was well aware of Frege's published writings because they were well known among the Poles during the early decades of twentieth century. According to Woleński (2004), Frege's ideas were frequently discussed by Łukasiewicz and Leśniewski. Woleński even remarks that '[i]n fact, Leśniewski's… work on definitions… is a continuation and extension of Frege's work' [45]. On Woleński's view, the main difference between Frege and Leśniewski is that while the former required non-creativity, the latter allowed creative definitions within his systems and thus, to avoid trouble, needed exact rules to govern them.

---

[25] In comparison with our discussion on Suppes (1957) in Sect. 6.3 it seems that Nemesszeghys deny the claim (b) but affirm claim (a) presented there.

[26] The Nemesszeghys give the citation in German and thank Owen Le Blanc for pointing it out to them. We will give it in English. The same reference (without a page number) is in Jurcic (1987, 198). He states that 'Leśniewski (1929a) first formulated the rules of definition and the requirements of eliminability and noncreativity.'

To whom the standard conditions finally will be attributed remains to be decided since we still know too little of the overall picture. It is quite possible that as potential *desiderata* on good definitions they formed ideas that were "in the air", and that they were, hence, common property.

To sum up, even though the folklore about Leśniewski and the requirements is dying away in the tradition following Belnap (1993), one can still run into people attributing the restrictions to Leśniewski. It seems also that the argument in Nemesszeghy and Nemesszeghy (1977) has not been contested before. This paper is meant as a *coup de grâce* to all such allegations. As we have indicated above, we will proceed through several steps. First we will explain why definitions in Leśniewski's systems are creative. Then we will look at Leśniewski's rules of definitions, and explain what they actually said: we will show that eliminability and conservativeness as the criteria (or a measure) of definitions are not to be found there. We will argue that, at least on Polish grounds, the study of these criteria should be credited to Łukasiewicz and Ajdukiewicz. Finally, we will reconsider Nemesszeghys' claim.

6.6 The Creativity of Leśniewski's Definitions

Now we will turn to Leśniewski's systems, and explain why some definitions in these systems are creative. To start with, definitions for Leśniewski are not meta-linguistic abbreviations; rather, he treats them as axioms (of a specific kind) formulated in the language of the system itself. The introduction of these definitions is to be governed by what he calls 'rules of definitions'.

Since our goal is only to explain why definitions in Ontology are creative, we can put many technical details aside and look at a rather simple example of definitions of name constants. Recall that the rule of definition for name constants says that a new constant γ can be introduced by means of a formula:

$$\forall a\,[a\,\varepsilon\,\gamma \equiv a\,\varepsilon\,a \wedge \phi(a)] \tag{6.1}$$

where $\phi(a)$ is a formula in the language of the system not containing any free variables other than a or other defined constants. (Strictly speaking, Leśniewski allowed defined constants to occur in defining conditions. He did this on the condition that the order in which constants have been introduced is maintained so that the definitions involved contain no circularity. We do not need this level of detail.)

The presence of $a\,\varepsilon\,a$ on the right-hand side might at first seem slightly surprising; the underlying idea here is that since the left-hand side assumes that a "is a singular term", the right-hand side has to do the same, and '$a\,\varepsilon\,a$' expresses exactly this statement (because no distinction between singular terms and other terms is built into the syntax, $a\,\varepsilon\,a$ has to be explicitly stated).

Thus, for instance, a formula that *prima facie* looks like a definition of Russell's class:

$$\forall a\,[a\,\varepsilon\,\lambda \equiv \neg a\,\varepsilon\,a] \tag{6.2}$$

is inadmissible as a definition; it is not a theorem of Ontology either, and it easily leads to contradiction since it entails $\lambda \, \varepsilon \, \lambda \equiv \neg \lambda \, \varepsilon \, \lambda$. Rather, the correct definition would be:

$$\forall a \, [a \, \varepsilon \, \lambda \equiv a \, \varepsilon \, a \wedge \neg a \, \varepsilon \, a] \tag{6.3}$$

which, since its right-hand side is a contradiction, entails:

$$\neg \exists a \, a \, \varepsilon \, \lambda, \tag{6.4}$$

which only says that nothing is λ (or that 'λ' doesn't name anything).

A closer examination reveals that (6.3) says that a is λ iff, first, a is an object, and second, a is not a. This avoids the contradiction because what we get when we substitute λ for a is:

$$\lambda \, \varepsilon \, \lambda \equiv \lambda \, \varepsilon \, \lambda \wedge \neg \lambda \, \varepsilon \, \lambda \tag{6.5}$$

Since the right-hand side of (6.5) is a straightforward contradiction, we can simply derive the negation of the left-hand side:

$$\neg \lambda \, \varepsilon \, \lambda \tag{6.6}$$

This, however, does not allow us to infer that $\lambda \, \varepsilon \, \lambda$.

By existential generalization, (6.4) entails that there is an empty name:

$$\exists b \, \neg \exists a \, a \, \varepsilon \, b. \tag{6.7}$$

This consequence, however, essentially relies on the definition of λ and is not provable in a system obtained by deleting the rule of definitions.

To see more clearly why the definition is creative, the following comparison would be useful. Consider a system which instead of the rule of definition contains what we may call *definitional comprehension*: for any $\phi(a)$ which satisfies the same conditions as those put on defining conditions occurring in definitions the following is an axiom:[27]

$$\exists b \, \forall a \, [a \, \varepsilon \, b \equiv a \, \varepsilon \, a \wedge \phi(a)] \tag{6.8}$$

(Similar additions are to be made for other types of definitions.) If we were to add to this system with definitional comprehension a definition (or any number of definitions) formulated in accordance with Leśniewski's schema (6.1), the obtained definitional extension would be non-creative.[28]

---

[27] Observe that by adding axioms of the form $\exists b \, \forall a \, (a \, \varepsilon \, b \equiv \phi(a))$ we would be able to derive the Russellian contradiction if we take $\phi(a)$ to be $\neg a \, \varepsilon \, a$.

[28] See Stachniak (1981) for a proof of a theorem from which our claim follows.

Thus the reason why definitions are creative in the original system is that Ontology, as it stands, lacks definitional comprehension: whenever one wants to prove $\exists b \, \phi(b)$ one has to define a γ, prove $\phi(\gamma)$, and then use existential generalization.[29]

So in a definition-free system with comprehension all constant-free theorems of the version with definitions but without comprehension are provable. The converse holds, too.

6.7 What Did Leśniewski's Rules for Definitions Actually Do?

Leśniewski (1931b) set out to give precise rules for definitions for Łukasiewicz's axiomatization of classical propositional logic.[30] Here Leśniewski uses his so-called terminological explanations to describe syntactically the axioms and the admissible rules of inference, including the rules of definitions (Leśniewski's rules of definitions for various systems are described in earlier chapters of this book). This syntactical focus makes his line of pursuit essentially different from Ajdukiewicz's strategy (which is discussed in the next section).

Crucial for Leśniewski's approach to definitions is his Terminological Explanation XI where he states what shape a definition is supposed to have. This explanation boils down to the requirement that a definition should be a formula of the form:

$$\neg[(\gamma(\alpha_1, \ldots, \alpha_n) \to \phi(\alpha_1, \ldots, \alpha_n)) \to \neg(\phi(\alpha_1, \ldots, \alpha_n) \to \gamma(\alpha_1, \ldots, \alpha_n))],$$

which is just a roundabout way of using negation and implication (which are primitive in the system) to express the equivalence:

$$\gamma(\alpha_1, \ldots, \alpha_n) \equiv \phi(\alpha_1, \ldots, \alpha_n)$$

where γ is a constant symbol being defined, $\alpha_1, \ldots, \alpha_n$ are all different propositional variables, ϕ contains only primitive (or previously defined) symbols, and the formulas contain only the explicitly mentioned variables.[31]

To get a clear picture of what is going on here we need only to look further at the last (twelfth) terminological explanation and the conclusion in Leśniewski (1931b). Terminological Explanation XII says 'of [an] object A that it is a definition, relative to C if and only if A is a definition of some expression, relative to C, by means of some expression, and with respect to some expression' [647]. This, when translated from *Leśniewskese*, basically states that a formula is a definition at a certain stage of development of a system if there is an expression of which it is a correct definition at that stage.

---

[29] Stachniak (1981) gives a Henkin-style completeness proof for a variant of Ontology which contains definitional comprehension for all categories of constants.

[30] Leśniewski (1931b) is a summary of the lectures that Leśniewski gave in Warsaw in 1930–1931.

[31] In Leśniewskianese this does have its bells and whistles and sounds a bit more complicated.

Leśniewski concludes the paper with a claim that a formula can be added to the system only if it is a consequence by substitution of previously proven theses, or it is a consequence by detachment of previously proven theses, or it is a correctly added definition. In general, Leśniewski's terminological explanations are meant to define the syntactic relation of derivability. They do not *employ* the notion itself and, *a fortiori*, they say nothing about conservativeness (This point holds for his treatment of definitions for all systems he considers).

At this point, it should be clear what Leśniewski set out to do and what he achieved. Using a rather idiosyncratic semi-formalized metalanguage, without any reference to eliminability or non-creativity, he provided a meticulous description of what syntactic form definitions should have. Definitions that satisfy his rules have consistency and eliminability properties. Yet he presented no proofs to this end: consistency is only mentioned in passing when Leśniewski says that the reason behind his rules is to ensure the consistency of the system and eliminability is just an unmentioned side-effect of the rules. Ajdukiewicz, on the other hand, instead of focusing on the syntactical form of definitions, tries to work out a more general motivation for them, suggesting that the syntactic restrictions result from certain more general meta-theoretical requirements.

6.8 Łukasiewicz and Ajdukiewicz on Definitions

On the Polish ground we can find some remarks pertaining to general meta-theoretical constraints on definitions in Jan Łukasiewicz's work, and a rather elaborate discussion in Kazimierz Ajdukiewicz's lecture scripts.

Łukasiewicz (1929),[32] in a rather short passage about definitions in propositional logic (in his logic course materials) remarks that substitution of defined terms should preserve the truth-value of sentences, and that addition of definitions should not make new expressions formulated in the original language provable:

> Sharing the view of the authors of Principia Mathematica I hold that definitions are theoretically superfluous. If we have a theory in which definitions do not appear at all, nothing new should be obtainable in that theory after we introduce definitions. (Łukasiewicz 1929, 52)

Alas, he also remarks that these issues do not belong to a general course in logic, and does not elaborate.

He, however, does not credit himself with the formulation of these criteria. In his introduction to this script he explicitly lists what he thinks the results he can claim are. He mentions (i) his bracket-free notation for propositional logic and Aristotle's syllogistic, (ii) his axiomatization of propositional logic, (iii) his way of writing down proofs in the systems in question and some of the proofs, (iv) his remarks about deduction (which do not pertain to definitions), (v) his systems of many-valued logics, (vi)

---

[32] As we mentioned in Sect. 6.3, Łukasiewicz (1929) was later published in English as Łukasiewicz (1963).

his completeness proof for propositional logic, (vii) his axiomatization of Aristotle's syllogistic, and (viii) some historical remarks about Aristotle, Stoics, Frege, Origen and Sextus. He notes that he could also claim (ix) his consistency proof for propositional logic and (x) his style of independence proofs, but those were independently invented by Post and Bernays. Most notably, he does not mention the restrictions on definitions, and he explicitly remarks: "Apart from the above-mentioned points, whatever can be found in the lectures, is not my property." (Łukasiewicz 1929, vii). He also observes that he owes a lot to discussions with his colleagues and their students and that there are many results he simply cannot correctly attribute.

Some new light can be cast on the history of the conditions when we look a few years earlier at the reports from the meetings of the Polish Philosophical Society that can be found in the 1928–1929 volume of *Ruch Filozoficzny* (*Philosophical Movement*). Łukasiewicz (1928a) describes a talk he gave at a plenary session of the Society on March 24, 1928, titled *The role of definitions in deductive systems*. There, he opposes to the idea that definitions should be interpreted as theorems of a given system and suggests that they rather should be interpreted meta-linguistically as abbreviations. The reason he presents is that if the former path is chosen, new theorems formulated in the language devoid of definitions can become provable. This, however, is not the whole story.

Only a few weeks before this talk, Łukasiewicz presented a more elaborate lecture in the Logic Section of the Society, titled "About definitions in theories of deduction".[33] Although Łukasiewicz does not say this, it is quite possible that he is attacking the views of Leśniewski who having criticized the meta-theoretical treatment of theorems and definitions in *Principia Mathematica*[34] decided to treat definitions intra-theoretically. Łukasiewicz starts off by presenting the opposition between two ways of interpreting definitions—as meta-theoretical abbreviations introduced by means of rules, and as theorems formulated within the system. Then he sketches two examples of creative definitions (in a propositional language) formulated using the latter, intra-theoretical approach (the creativity of one of them has been proven by Wajsberg, and of the other by Łukasiewicz). Łukasiewicz's main point is that we should interpret definitions meta-theoretically since their creativity, which clearly can take place if definitions are interpreted intra-theoretically, is undesirable.

Łukasiewicz mentions then that Professor Leśniewski participated in the discussion insisting that 'in [his] Ontology definitions lead to theses independent of the axioms; this is not a vice; quite the contrary: if one adds definitions, creative is exactly what they should be.' [178]

A few points seem worth bringing up. First, Łukasiewicz's method of finding independence proofs plays a key role here (Roughly, for a propositional language the strategy is that if one wants to show that given a certain Hilbert-style proof system a formula ϕ is independent of a premise set Γ, one has to find some many-valued characterization of the connectives occurring in the language, on which all

---

[33] A report on this talk Łukasiewicz (1928b) (written by Łukasiewicz himself) appeared also in the same volume as the other report.

[34] See Urbaniak (2008b, 85–92) for details.

assumptions in Γ have one of the chosen values, inference rules preserve chosen value(s), and yet ϕ does not have a chosen value.). Indeed, to prove that a definition is creative in a certain system one not only has to establish that with this definition one can prove a certain formula, but also that this formula is not provable in the system itself, i.e. that it is *independent* of the original axioms.

Second, it is still rather unclear what role creativity plays in arguments against the object-language treatment of definitions. What Łukasiewicz seems to have proven is that if certain definitions are accepted as theorems, they are creative. But even if one values non-creativity, to turn this into an argument against the object-language treatment of definitions, one also has to show that once one switches to the meta-linguistic treatment of definitions, non-creativity vanishes. This however, *prima facie*, seems unlikely: if you can prove a new χ with a theorem $\phi \equiv \psi$, you are also able to prove the same χ with a meta-theoretical rule that captures this equivalence. Thus it appears probable that (at least in some contexts) the distinction between the creative and the conservative cuts across the one between the intra-theoretic and the meta-theoretical.

Third, the debate emphasizing the importance of non-creativity seems to stem from Leśniewski's view of definitions as theorems, even if it was not Leśniewski who formulated the requirement: this at least partly explains what Leśniewski's impact on these matters was and why his name came to be connected with the standard requirements. Łukasiewicz's report also constitutes evidence for the claim that once Leśniewski was faced with the non-creativity requirement he rejected it.

Now we may turn to Ajdukiewicz's contribution. Ajdukiewicz discusses translatability (which is, *mutatis mutandis*, the same as eliminability) and consistency requirements in a lecture script (in Polish) which dates back to 1928. Ajdukiewicz used these notes when he was teaching in Warsaw. (Those parts of Ajdukiewicz (1928) that pertain to definitions have been published in Polish in 1960.)

Furthermore, translatability, consistency and conservativeness requirements are discussed in a paper Ajdukiewicz (1936) gave later in Paris.[35] As it will turn out by

---

[35] In the 1956 edition of Tarski's *Logic, Semantics, and Metamathematics*, in the translation of Tarski (1934), 'consistency and re-translatability' are mentioned as 'the conditions for a correct definition' [307, n. 3]. At the same page of the 1983 edition the criteria are 'non-creativity and eliminability'. This puzzling fact is noticed by Hodges (2008). Paolo Mancosu pointed out an interesting passage concerning the second edition, written by Tarski in his correspondence with Corcoran:

> Replace "re-translatability" by "non-creativity". [I do not explain the meaning of the term "non-creativity" for the same reason why I did not explain before the meaning of "re-translatability". I have never intended to make LSM a self-contained work. By the way, re-translatability is a stronger property than "non-creativity". In old times it was frequently used in discussing definitions. It seems that now "non-creativity" is more fashionable.]

Tarski might have been alluding to the work done in the theory of definability back in the 1960s and 1970s. One of the results was that (supposing consistency) the eliminability of a term is a necessary and sufficient condition for its explicit definability within the given theory, whereas conservativeness is only sufficient: some weaker forms of definability imply the latter property as well (See Rantala 1977, 179–185). But for some reason the 1983 footnote does not read 'consistency and non-creativity', which would have been the weaker criteria.

the end of this section, Ajdukiewicz not only mentions these rules, but also provides some proofs concerning them.

In the introduction to a collection of his papers, around 30 years after having written the script, Ajdukiewicz (1960, v–vi) explains his motivation behind the 1928 and 1936 papers by saying that he tried to understand what the ultimate goal of structural rules for definitions are. He contrast his own approach to Leśniewski's rules which were purely syntactical.

Ajdukiewicz observes what we have already learned in the previous sections: Leśniewski was not concerned with general conditions on definitions formulated in terms of derivability, but rather with the project of defining derivability in terms of syntactic relations, which includes a purely syntactic description of what a definition should look like. Ajdukiewicz himself, however, was after more general conditions: those directly related to the purpose a definition should serve in a system.

In the 1928 script, which is titled *The main principles of methodology of sciences and formal logic (Główne zasady metodologii nauk i logiki formalnej)*, Ajdukiewicz first introduces the notion of being a meaningful expression relative to (a background theory consisting of) sentences Z: An expression W is, in this sense, meaningful if it contains only such constants which are equiform with one of the constants occurring in Z, and if its syntax obeys the syntax of Z. [Par. 12, p. 45].

The notion of a meaningful expression is used in paragraph 19 titled *Rules of Definitions (Dyrektywy definiowania)* to introduce the requirements of translatability and consistency, pretty much as we know them. Say we extend the language of Z, J_Z, i.e. 'the set of sentences meaningful relative to Z', to a new language J_{Z+D} by using a definition D to introduce an expression δ, which is new relative to Z. The first condition Ajdukiewicz introduces is translatability which requires that any sentence in J_{Z+D} should (*modulo* accepted inference rules and theory $Z + D$) be inferentially equivalent to a sentence in J_Z. The second condition he introduces is consistency: if Z was consistent (relative to given inference rules which are kept fixed) then $Z + D$ also has to be consistent. (Ajdukiewicz 1928, 46–47).

Non-creativity, even though not mentioned in 1928, receives attention only a few years later. (1936, 244), after pretty much repeating his previous formulation of translatability and consistency requirements, remarks that...

> ...Often, but not always, one also wishes the rules of definitions to exclude creative definitions. A definition is called creative on the grounds of a certain language, if from the theses of that language according to the rules of deduction by means of that definition it is possible to derive a sentence of that language (and so, not containing the defined term), which cannot be derived without that definition.

In other words he claims that sometimes one requires also that no formulas of the old language which were not theorems of the original theory become derivable in the theory obtained by adding a definition. This is exactly the non-creativity mentioned e.g. by Suppes (1957, 153),[36] by Łukasiewicz (see above), and already by Frege:

> In fact it is not possible to prove something new from a definition alone that would be unprovable without it. When something that looks like a definition really makes it possible

---

[36] See the quotation in Sect. 6.3.

to prove something which could not be proved before, then it is no mere definition but must conceal something which would have either to be proved as a theorem or accepted as an axiom. (Frege 1914, 208)

Although it is unlikely that Ajdukiewicz independently reinvented the conditions of eliminability and conservativeness (as they are called nowadays), what sets him apart from his predecessors and contemporaries is his genuinely meta-meta-theoretical treatment of these criteria.[37] He, for instance, not only mentions non-creativity in (1936, 245–246), but also attempts to find formal conditions whose satisfaction by a system guarantees the non-creativity of definitions within the system. He assumes the consistency requirement and argues that if the rules of inference cannot distinguish between constants, definitions are non-creative on a rather straightforward condition of irrelevancy of the defined terms for the language's inference rules (i.e. that derivability is preserved under uniform substitution).

Given this assumption of "irrelevancy", the argument that the consistency requirement is (in such a setting) sufficient for non-creativity becomes rather straightforward. Suppose a definition D of a word W leads to inconsistency with premises Z, where formulas in Z do not contain W. Then we can with a substitution of D, which instead of W contains an expression of the same category but occurring already in Z, derive a contradiction just from Z as well. So for D to satisfy the consistency requirement it is sufficient that Z derives an instance of the definition which is constructed within the language of Z. Of course, if the consistency requirement is not satisfied then definitions will be creative, allowing for the derivation of $\bot$ which was underivable in the system devoid of definitions.

Ajdukiewicz observes that a similar reasoning applies to creativity. If a W-free formula ϕ is to be derivable from D with Z, then if Z already proves a W-free instance of D then Z also proves ϕ.

It should be clear how this applies to the creativity of Leśniewski's definitions. The condition sufficient for non-creativity discussed by Ajdukiewicz requires that no constants should be distinguished in the system. In QNL_{df} it is clearly violated: it is possible to derive expression (6.3) from Sect. 6.6, that is:

$$\forall a\, [a\, \varepsilon\, \lambda \equiv a\, \varepsilon\, a \wedge \neg a\, \varepsilon\, a]$$

but (without the definition itself) it is neither possible to derive an instance of it:

$$\forall a\, [a\, \varepsilon\, b \equiv a\, \varepsilon\, a \wedge \neg a\, \varepsilon\, a]$$

nor its existential generalization:

---

[37] For example, we do not find such approach in Frege (1914), Łukasiewicz (1929) or Suppes (1957). More recently, however, Ajdukiewicz's interest in the criteria reappear independently in e.g. Belnap (1993) and Došen and Schroeder-Heister (1985). Yet, since the history of the standard theory remains still unwritten, the study of little-known sources can, as our discussion here demonstrates, reveal some surprises.

$$\exists b \, \forall a \, [a \, \varepsilon \, b \equiv a \, \varepsilon \, a \wedge \neg a \, \varepsilon \, a].$$

In a sense, the whole point of introducing a definition in QNL_{df} is to distinguish one constant and to be able to prove about it something not provable about any other constant. Also, in a way, it is the idea that all we need is instances of definitions that stands behind the move to QNL_{com} and the conservativeness of definitions in that system.

Later Ajdukiewicz (1958) presented a sketch of what he called 'a general theory of definitions' which would treat about all types of definitions: real, nominal, and conventional, and which would reveal the logical relations between these classes of definitions. On both of these tasks the criteria of translatability and non-creativity (latter in a disguise of a need to proof existence) are utilized.[38]

To sum up, Łukasiewicz and Ajdukiewicz in the 1920s and 1930s both mention and use the consistency, translatability, and conservativeness requirements for definitions. Ajdukiewicz studied them and indicated one source of the creativity of definitions in some systems: the fact that certain constants are in such systems, so to speak, distinguished.

6.9 Remarks

We have tried to provide answers to three sets of problems: (i) where did the idea that it was Leśniewski who introduced the restrictions of eliminability and conservativeness originate and why has this impression been around for so long? (ii) is the conviction true at all, i.e. was it really Leśniewski who established these requirements and, if not, what was his actual stance on definitions? (iii) if not Leśniewski then who in fact presented and studied the meta-theoretical criteria of good definitions in Poland?

Let us briefly summarize our findings. The myth about Leśniewski's role has died hard because:

1. It is general knowledge that Leśniewski's works on definitions are of importance,
2. Leśniewski is often praised for having introduced precise rules for definitions by authors who leave the nature of those rules obscure (e.g. Luschei 1962), or even misrepresent them (e.g. Suppes 1957), and the influence of some of these texts has been tremendous,
3. the accessibility of Leśniewski's works has been low: the English translation of his collected works dates back only to 1991[39] and given his idiosyncratic style and the complexity of his formulations the translations are not much easier to read (even for non-Polish and non-German speakers),

---

[38] Ajdukiewicz's general theory could be more fruitful theoretical basis for a practical account of definitions used in computer science and network technologies than, for instance, that of Robinson (1954), whose system is developed further in Cregan (2005).

[39] Some of Leśniewski's texts have been translated before, most notably Leśniewski (1931b) in the important collection of papers by Polish logicians: (McCall 1967).

4. further, the availability of the relevant works by Łukasiewicz and by Ajdukiewicz is even lower, especially for non-Polish readers.[40]

With high plausibility we can state that it was *not* Leśniewski who introduced translatability, consistency and conservativeness requirements for definitions; and, for certain, he did not study them:

5. Leśniewski's use of creative definitions appears essential to his systems,
6. Leśniewski's rules of definitions are concerned only with the syntactic form of definitions as admitted in particular systems,
7. Leśniewski does not bring up eliminability and non-creativity requirements even as potential desiderata of good definitions anywhere in his writings (not even in those bits which are devoted to definitions),
8. Ajdukiewicz, who was familiar with Leśniewski's work and knew him personally, also states that Leśniewski has formulated no meta-theoretical requirements for definitions (and this fact was, according to Rickey (1975b), confirmed by B. Sobociński as well).

Most importantly, we have found good reasons to believe that it was Łukasiewicz who brought the issue up, and Ajdukiewicz who was first to provide a meta-theoretical study of how the criteria of conservativeness affect the definitions, at least on the Polish ground:

9. Łukasiewicz in 1928 uses the conservativeness requirement to argue against the intra-theoretic treatment of definitions and presents the criterion in his lecture script (although he does not credit himself with its formulation),
10 Ajdukiewicz describes the translatability and consistency requirements explicitly in 1928,
11. Ajdukiewicz mentions and studies, with respect to some other meta-theoretical conditions, conservative definitions in 1936,
12. Ajdukiewicz seems to be crediting *not* Leśniewski but himself with the claim that the goal of Leśniewski's syntactic restrictions is to satisfy the meta-theoretical requirements Ajdukiewicz mentioned in 1928.

There still are many open questions pertaining to the history of the standard account of definitions (as well as to the history of the non-standard account of creative and/or circular definitions). For instance, what happened during the 20 year gap between the works of Frege, Peano, Russell and Whitehead, and the works of Łukasiewicz, Leśniewski and Ajdukiewicz? How does Dubislav fit into the picture? There seems to be room for a more extensive study of Peano's group as well.

As is clear from our discussion, the study of the Polish golden age of philosophy and logic, i.e. the time between Kazimierz Twardowski's (1866–1938) appointment as a professor at Lvov in the end of nineteenth century and the Second World War, might be sometimes surprising.

Even if some pieces of the puzzle are still missing, it seems we can now at least conclude: The myth about Leśniewski and definitions is (almost) definitely busted.

---

[40] The best source for translations of their seminal works in English is McCall (1967).

References

Abelson, R. (1967). Definition. In P. Edwards (Ed.), *The encyclopedia of philosophy* (Vol. 2, pp. 314–324). New York: Macmillan & The Free Press.

Ajdukiewicz, K. (1928). Definicja. [Parts published in Język i Poznanie. Tom I. Wybór Pism z Lat 1920–1939, Warszawa 1960 and 1985, pp. 44–61].

Ajdukiewicz, K. (1936). Die Definitionen. In Actes du I Congrés International de Philosophie Scientifique, pp. 1–6. Hermann, Paris. (The Polish version translated by Franciszek Zeidler published in Język i Poznanie. Tom I. Wybór Pism z Lat 1920–1939, Warszawa 1960 and 1985, pp. 243–248).

Ajdukiewicz, K. (1960). Przedmowa. In Język i Poznanie. Tom I. Wybór Pism z Lat 1920–1939, pp. v-viii. PWN, Warszawa (Introduction to the collection).

Ajdukiewicz, K. (1958). Three concepts of definition. *Logique et Analyse, 1*(3–4), 115–126.

Beck, L. W. (1956). Kant's theory of definition. *The Philosophical Review, 65*(2), 179–191.

Belnap, N. D. (1962). Tonk, plonk and plink. *Analysis, 22*(6), 130–134.

Belnap, N. (1993). On rigorous definitions. *Philosophical Studies, 72*(2–3), 115–146.

Betti, A. (2008). Polish axiomatics and its truth - on Tarski's Leśniewskian background and the ajdukiewicz connection. *In Pattnew, 2008,* 44–71.

Church, A. (1956). Introduction to mathematical logic (Vol. 1). Princeton: Princeton University Press.

Cregan, A. M. (2005). Towards a science of definition. *AOW '05: Proceedings of the 2005 Australasian Ontology Workshop* (pp. 25–32). Australian Computer Society: Darlinghurst, Australia.

Došen, K., & Schroeder-Heister, P. (1985). Conservativeness and uniqueness. *Theoria, 51,* 159–173.

Dubislav, W. (1981). Die Definition (4th ed.). Hamburg: Felix Meiner Verlag (Unchanged from the 3rd ed. of 1931).

Dudman, V. H. (1973). Frege on definition. *Mind, 82*(328), 609–610.

Fetzer, J. H., Shatz, D., & Schlesinger, G. N. (Eds.). (1991). *Definitions and definability: philosophical perspectives, volume 216 of synthese library.* Dordrecht: Kluwer.

Frege, G. (1879). Begriffsschrift: eine der arithmetischen nachgebildete Formelsprache des reinen Denkens. L. Nebert, Halle. (Page numbers refer to the translation in Heijfrom (1967), pp. 1–82).

Frege, G. (1914). Logic in mathematics. (A lecture note, German original in Hermes et al. (1969), page numbers refer to the translation in (Hermes et al. 1979, 203–250)).

Galilei, G. (2001). Dialogues Concerning Two New Sciences. Norwich: William Andrew Publishing (Translation by Henry Crew and Alfonso de Salvio, Electronic publication, retrieved from www.knovel.com).

Grattan-Guinness, I. (2000). *The Search for mathematical roots, 1870–1940: Logics.* Set theories and the foundations of mathematics from cantor through russell to Gödel. Princeton: Princeton University Press.

Gupta, A. (2009). Definitions. In E. N.Zalta (Ed.), The Stanford Encyclopedia of Philosophy. Spring 2009 edition.

Hacking, I. (1979). What is logic? *The Journal of Philosophy, 76*(6), 285–319.

Hodges, W. (2008). *Tarski's theory of definition. In Pattnew, 2008,* 94–132.

Horty, J. (2007). *Frege on definitions: a case study of semantic content.* Oxford: Oxford University Press.

Jordan, Z. (1945). The development of mathematical logic and of logical positivism in poland between the two wars. Polish Science and Learning. London: Oxford University Press (Partly republished in (McCall, 346–397)).

Jurcic, J. (1987). On defining sentential connectives. *Notre Dame Journal of Formal Logic, 28*(2), 189–199.

Kelley, J. L. (1955). *General topology.* Princeton: The University Series in Higher Mathematics. D. van Nostrand Company.

Kennedy, H. C. (1973). What russell learned from peano. *Notre Dame Journal of Formal Logic, 14*(3), 367–372.

Koj, L. (1987). *Walter Dubislav und Kazimierz Ajdukiewicz über die Definition* (pp. 93–116). Grammar and Rhetoric, VI: Studies in Logic.

Leśniewski, S. (1929a). Grundzüge eines neuen Systems der Grundlagen der Mathematik §1-11. Fundamenta Mathematicae, 14,1–81. [Fundamentals of a new system of the foundation of mathematics, §1-11, (Leśniewski, 1991, 410–605)].

Leśniewski, S. (1931a). O podstawach matematyki, Rozdział X: Aksjomatyka 'ogólnej teorji mnogości pochodząca z r. 1921. Rozdział XI: O zdaniach 'jednostkowych' typu 'Aεb'. Przegląd Filozoficzny.

Leśniewski, S. (1931b). Über Definitionen in der sogenannten Theorie der Deduction. Sprawozdania z posiedzeń Towarzystwa Naukowego Warszawskiego, Wydział Nauk Matematyczno-Fizycznych, 24:289ñ309. [On definitions in the so-called theory of deduction, (Leśniewski, 1991,629–648].

Łukasiewicz, J. (1929). Elementy logiki matematycznej. [Reprinted in 1958 in Warsaw by Państwowe Wydawnictwo Naukowe and in 2008 in Poznań, Wydawnictwo Naukowe UAM].

Łukasiewicz, J. (1963). Elements of mathematical logic. International Series of Monographs in Pure and Applied Mathematics. Oxford: Pergamon Press (Translation of Elementy Logiki Matematycznej, PWN, Warszawa, 1958. Originally appeared as mineographed notes, (Łukasiewicz, 1929)).

Łukasiewicz, J. (1928). Posiedzenie naukowe sekcji logiki polskiego towarzystwa filozoficznego z dnia 18 lutego 1928, autoreferat. *Ruch Filozoficzny, 11*, 177–178.

Łukasiewicz, J. (1928). Plenarne posiedzenie naukowe z dnia 24 marca 1928, autoreferat. *Ruch Filozoficzny, 11*, 164.

Luschei, E. (1962). *The Logical systems of leśniewski.* Amsterdam: North-Holland.

Mates, B. (1972). *Elementary logic* (2nd ed.). New York: Oxford University Press.

McCall, S. (Ed.). (1967). *Polish Logic 1920–1939.* Oxford: Clarendon press.

Mill, J. S. (1869). *A system of logic, ratiocinative and inductive: being a connected view of the principles of evidence and the methods of scientific investigation.* New York: Harper.

Mostowski, A. (1948). Logika Matematyczna. Kurs Uniwersytecki [Mathematical logic. A university course]. Seminarium Matematyczne Uniwersytetu Wrocławskiego.

Nemesszeghy, E. Z., & Nemesszeghy, E. A. (1977). On strongly creative definitions: a reply to Rickey, V.F. Logique et Analyse, 20(77–78),111–115.

Nemesszeghy, E. Z., & Nemesszeghy, E. A. (1971). Is $(p \supset q) = (\sim p \vee q)$ Df. a proper definition in the system of principia mathematica? *Mind, 80*(318), 282–283.

Nemesszeghy, E. Z., & Nemesszeghy, E. A. (1973). On the creative role of the definition $(p \supset q) = (\sim p \vee q)$ df. in the system of principia: reply to V. H. Dudman (i) and R. Black (ii). *Mind, 82*(328), 613–616.

Padoa, A. (1900). Logical introduction to any deductive theory. In Heijfrom 1967, pp. 119–123. (Translation of part of 'Essai d'une théorie algébrique des nombres entiers', précédé d'une introduction logique à une théorie déductive quelconque', originally published in Bibliothèque du Congrès International de Philosophie, Vol. 3, Armand Colin, Paris).

Pascal, B. (1814a). De l'art de persuader. In Pascal (1814c), pp. 153–173.

Pascal, B. (1814b). De l'esprit géométrique. In Pascal (1814c), pp. 123–152.

Prior, A. N. (1960). The runabout inference-ticket. *Analysis, 21*(2), 38–39.

Rantala, V. (1977). *Aspects of definability* (Vol. 29)., of acta philosphica fennica Amsterdam: North-Holland Publishing Company.

Rickey, V. F. (1975b). On creative definitions in the principia mathematica. *Logique et Analyse, 18*(69–70), 175–182.

Rickey, V. F. (1975a). Creative definitions in propositional calculi. *Notre Dame Journal of Formal Logic, 16*(2), 273–294.

Robinson, R. (1954). *Definition.* Oxford: Oxford University Press.

Sager, J. C. (Ed.). (2000). *Essays on definition.* Amsterdam: Terminology and Lexiography Research and Practice. John Benjamins Publishing Co.

Shieh, S. (2008). Frege on definitions. philosophy. *Compass, 3*(5), 992–1012.

Simons, P. (2008b). Stanisław Leśniewski. E. N.Zalta (Ed.), Stanford Encyclopedia of Philosophy. Fall 2008 edition.

Słupecki, J. (1955). St Leśniewski's calculus of names. *Studia Logica, 3*, 7–72.

Stachniak, Z. (1981). *Introduction to model theory for lesniewski's ontology.* Wrocław: Wydawnictwo Uniwersytetu Wrocławskiego.

Suppes, P., & (1957). Introduction to Logic. New York: Van Nostrand Reinhold (Republished unabridged,. (1999). *Dover Publications.* New York): Mineola.

Tarski, A. (1934). Z badań metodologicznych nad definiowalnością terminów. Przegląd Filozoficzny, 37, 438–460. [Translated as "Some methodological investigations on the definability of concepts" in (Tarski, 1956, 298–319)].

Tarski, A. (1941). Introduction to Logic and to the Methodology of Deductive Sciences. Oxford: Oxford University Press [Original Polish edition 1936, translated by Olaf Helmer from the German edition, Tarski (1937)].

Tarski, A. (1956). Logic, semantics, metamathematics. Oxford: Claredon Press [Translation by J. H. Woodger].

Tarski, A. (1983). Logic, semantics, metamathematics (2nd ed.). Indianapolis: Hackett Publishing Company [Edited by John Corcoran, translation by J. H. Woodger].

Tarski, A. (1937). *Einführung in die mathematische logik und in die methodologie der mathematik.* Wien: Julius Springer.

Urbaniak, R. (2008b). Leśniewski's systems of logic and mereology. History and Re-evaluation. PhD thesis, University of Calgary [Available at https://dspace.ucalgary.ca/handle/1880/46697]

Urbaniak, R., & Hämäri, K. (2012). Busting a myth about Leśniewski and definitions. *History and Philosophy of Logic, 33*(2), 159–189.

Woleński, J. (1985). Filozoficzna Szkoła Lwowsko-Warszawska. PWN, Warszawa. [Translated as Logic and philosophy in the Lvov-Warsaw school, Reidel, Dordrecht, 1989].

Woleński, J. (2004). The reception of frege in poland. *History and Philosophy of Logic, 25*(1), 37–51.

Chapter 7
Sets Revisited

Abstract I discuss various attempts to emulate standard set theory within the framework of Leśniewski's system: the Leśniewski–Sobociński strategy of distinguishing between collective and distributive totalities (a heap of stones is a collective totality; a set of chairs is not) I show the strategy, as used by Sobociński, leads to a few paradoxes. Further, I argue that using the means available within Ontology itself does not result in a theory strong enough to mimic set theory. I then describe Słupecki's *generalized mereology* and argue that it is too different from set theory to be able to play a foundational role.

7.1 Distributive Classes

Intuitively, there is in English a name-forming functor of one name argument. This functor forms with its argument a name of the class of all and only those objects which are named by its argument. It is *'the class of'* (henceforth *'cl'*). So, for instance, we speak of the class of integers, the class of bears, the class of contradictory objects (presumably, it is the empty class), and so on.

While I already covered the mereological notion of a class in this book, the functor I now have in mind is *distributive*, as opposed to the *collective* class-forming functor. In the collective sense the class of a's is the result of collecting them into one mereological whole. In the distributive sense, the property of being an a distributes over the elements of the class of a's, so that whatever belongs to the class of a's is an a. In this distributive sense, the following sentences seem compelling:

(7.1) My dog's leg is not an element of the class of dogs.

(7.2) The class of integers is not an element of the class of integers.

(7.3) The class of the class of elephants is not the same as the class of elephants.

(7.4) My dog is an element of the class of dogs. The class of dogs is an element of the class of classes of animals But my dog is not an element of the class of classes of animals.

R. Urbaniak, *Leśniewski's Systems of Logic and Foundations of Mathematics*,
Trends in Logic 37, DOI: 10.1007/978-3-319-00482-2_7,
© Springer International Publishing Switzerland 2014

These are examples of sentences true in the distributive reading and false in the collective interpretation.[1] The first sentence indicates that the notion of distributive element is different from parthood: a part of an element of a class does not have to be an element of this class. Distributively speaking, the relation of being an element is not reflexive. It is also sensitive to iterations and intransitive.

In this distributive sense I will take for granted the following:

$$\forall a, b[b \ \varepsilon \ el(a) \equiv \exists c \ (a \varepsilon cl(c) \wedge b \varepsilon c)] \tag{7.5}$$

where $el(a)$ is read as *an element of a*. It states that an object b is an element of a if and only if there is a possible name c such that a is the class of all and only those objects that are c and b is c. (7.5) will constitute the hard core of most things which follow in this chapter.

It is not clear how Leśniewski would account for this sort of distributive intuitions. Perhaps, Leśniewski would say that the only irreducible sense of the word 'class' is that formalized by Mereology. After all, he employed mereological intuitions to reject (2.72) (see page xx of this book).[2] Perhaps, he would concede that there are two interpretations of the notion of a class involved. At least, this is the way that Leśniewski's student, Sobociński (1949), interpreted the situation. We will take a closer look at what the latter has to say in Sect. 7.4, but before we do that, let us revisit Russell's paradox once again, this time within the formal framework.

7.2 Russell's Paradox Again

The first attempt to extend (7.5) to a substantial theory may consist in adding the following axioms:

$$\forall a \exists b \ (b \varepsilon cl(a)) \tag{7.6}$$

---

[1] It is important to emphasize that the question I am concerned with is **not** whether there is a way of introducing into the language of Ontology a formal counterpart of the expression 'the class of' which would mimic correctly Leśniewski's understanding of this term (which was quite unusual), but rather how the expression 'the class of' in the nowadays most common, set-theoretic use can be imitated in the language of Leśniewski's systems. Thus, even if Leśniewski would not share the intuitions behind the examples above, it does not matter. The point is that the sense of 'the class of' in which the above examples come out true seems to be a legitimate and widespread understanding of the expression in question and Mereology falls short of accounting for this plausibility.

[2] There is an intriguing comment on Leśniewski's philosophical method made by Twardowski: "In general, those who follow Leśniewski, very arbitrarily demand an analysis where they find it convenient; however, whenever someone demand an analysis where it is inconvenient, they refer to intuition. And when the opponent in the discussion tries at some point to refer to intuition as well, they respond: "We cannot understand what you claim to be intuitively given". (*K. Twardowski's Diary*, ms. 2407/3)". [The quote comes from Kazimerz Twardowski's archive, located in the library of the Institute of Philosophy and Sociology of the Polish Academy of Sciences, Warsaw. The reference is to signatures in this collection.]

$$\forall a, b, c, d[a\varepsilon cl(c) \land cl(d) \land b\varepsilon c \to b\varepsilon d] \tag{7.7}$$

(7.6) is full comprehension and (7.7) says that a class can be determined only by names of the same extension.

Unfortunately, this will not work. We can easily define a name $\star$ by:

$$\forall a[a\varepsilon\star \equiv a\varepsilon a \land \forall b(a\varepsilon cl(b) \to \neg a\varepsilon b)] \tag{7.8}$$

The existence of $cl(\star)$ is guaranteed by (7.6). The class is something similar to the Russell's class. (Although note that it does not even require that a be a class in order to be $\star$.) Clearly, (7.6), (7.7) with (7.8) imply contradiction.[3]

Theorem 7.1 (Sobociński) (7.6) (7.7) (7.8) $\vdash \bot$.

To see why this holds consider the following proof:

1	$a\varepsilon\star \to (a\varepsilon cl(\star) \to \neg a\varepsilon\star)$	Def. (7.8)
2	$\neg a\varepsilon\star \to (a\varepsilon cl(\star) \to \neg a\varepsilon\star)$	CL
3	$a\varepsilon cl(\star) \to \neg a\varepsilon\star$	CL: 1, 2
4	$a\varepsilon cl(b) \land a\varepsilon cl(\star) \land a\varepsilon b \to a\varepsilon\star$	(7.7)
5	$a\varepsilon cl(b) \land a\varepsilon cl(\star) \to \neg a\varepsilon b$	CL: 3, 4
6	$\quad a\varepsilon cl(\star)$	Assumption
7	$\quad a\varepsilon a$	Ontology: 6
8	$\quad \forall b (a\varepsilon cl(b) \to \neg a\varepsilon b)$	CL$\forall$: 5, 6
9	$\quad a\varepsilon\star$	Def. (7.8): 7, 8
10	$a\varepsilon cl(\star) \to a\varepsilon\star$	6$\Rightarrow$9
11	$\neg\exists a\, a\varepsilon cl(\star)$	Ontology: 3, 10
12	$\exists a\, a\varepsilon cl(\star)$	(7.6)
13	$\bot$	10, 11

---

[3] The first four theorems in this chapter have been proven in a rather tedious manner by Sobociński (who credits Leśniewski with the proofs). I give simplified and streamlined proof sketches based on Sobociński's proofs.

Maybe something is wrong with (7.6)? Let us try to weaken this assumption to the following:

$$\forall a, b[b\varepsilon a \rightarrow \exists c c\varepsilon cl(a)] \tag{7.9}$$

This restricts comprehension to non-empty names. On Leśniewski's view, this might be more intuitive, because he did not believe in empty classes. However, the problem remains. Assume (7.9) and (7.7). These assumptions together with (7.8) lead to the conclusion that if an object is a class of b's, it is b itself:

$$\forall a, b[a\varepsilon cl(b) \rightarrow a\varepsilon b] \tag{7.10}$$

Theorem 7.2 (7.9), (7.7), (7.8) $\vdash$ (7.10).

Start the proof as above and continue from line 11:

1	$\forall a \, \neg a \, \varepsilon \star$	line 11
2	$a \, \varepsilon \, cl(b)$	Assumption
3	$a \, \varepsilon \, a$	Ontology: 2
4	$\neg a \, \varepsilon \star$	CL$\forall$: 1
5	$\exists c \, [a \, \varepsilon \, cl(c) \wedge a \, \varepsilon \, c]$	Def (7.8), 3, 4
6	$a \, \varepsilon \, cl(c) \wedge a \, \varepsilon \, c$	Assumption
7	$a \, \varepsilon \, cl(c) \wedge a \, \varepsilon \, cl(b) \wedge a \, \varepsilon \, c \rightarrow a \, \varepsilon \, b$	(7.7)
8	$a \, \varepsilon \, b$	CL: 2, 6
9	$a \, \varepsilon \, b$	$\exists$Elim: 5, 6$\Rightarrow$8
10	$\forall a \, [a \, \varepsilon \, cl(b) \rightarrow a \, \varepsilon \, b]$	

The problem is, (7.10) agrees neither with the distributive nor with the collective notion of a class. For neither the distributive class of natural numbers is a natural number itself nor the collective mereological fusions of all lizards in the world is a lizard.

There is even a worse problem. To see it, we first need the following lemma:

$$\forall a, b, c[a\varepsilon a \wedge c\varepsilon cl(a \cup b) \wedge b\varepsilon c \to a = b] \tag{7.11}$$

The proof is:

1	$a\,\varepsilon\,a \wedge c\,\varepsilon\,cl(a \cup b) \wedge b\,\varepsilon\,c$	Assumption
2	$b\,\varepsilon\,b$	Ontology: 1
3	$a\,\varepsilon\,a \cup b$	Ontology: 1
4	$\exists d\ d\,\varepsilon\,cl(b)$	Ontology: 2, (7.9)
5	$\quad d\,\varepsilon\,cl(b)$	Assumption
6	$\quad d\,\varepsilon\,b$	Ontology: 5, (7.10)
7	$\quad d\,\varepsilon\,c$	Ontology: 6, 1
8	$\quad d\,\varepsilon\,cl(a \cup b)$	Ontology 7, 1
9	$\quad d\,\varepsilon\,cl(a \cup b) \wedge d\,\varepsilon\,cl(b) \wedge a\,\varepsilon\,a \cup b \to a\,\varepsilon\,b$	(7.7)
10	$\quad a\,\varepsilon\,b$	CL: 8, 5, 3
11	$\quad a = b$	Ontology: 2, 10
12	$a = b$	$\exists$Elim: 4, 5$\Rightarrow$11

Now, it turns out that the current axioms entail that there exists at most one object:

$$\forall a, b[a\varepsilon a \wedge b\varepsilon b \to a\varepsilon b] \tag{7.12}$$

which seems even more problematic than (7.10).

Theorem 7.3 (Sobociński) (7.7), (7.8), (7.9) $\vdash$ (7.12).

Here is the proof:

1	$a\,\varepsilon\,a \wedge b\,\varepsilon\,b$	Assumption
2	$a\,\varepsilon\,a \cup b$	Ontology: 1
3	$\exists c\,c\,\varepsilon\,cl(a \cup b)$	Ontology 2, (7.9)
4	$c\,\varepsilon\,cl(a \cup b)$	Assumption
5	$c\,\varepsilon\,a \cup b$	Ontology, 4, (7.10)
6	$c\,\varepsilon\,a \vee c\,\varepsilon\,b$	Def.: 5
7	$a\,\varepsilon\,c \vee b\,\varepsilon\,c$	Ontology: 1, 6
8	$a\,\varepsilon\,c$	Assumption
9	$b\,\varepsilon\,b \wedge c\,\varepsilon\,cl(a \cup b) \wedge a\,\varepsilon\,c \to a = b$	(7.11)
10	$a = b$	CL:, 1, 4, 8, 10
11	$b\,\varepsilon\,c$	Assumption
12	$a\,\varepsilon\,a \wedge c\,\varepsilon\,cl(a \cup b) \wedge b\,\varepsilon\,c \to a = b$	(7.11)
13	$a = b$	CL: 1, 4, 11, 12
14	$a = b$	CL: 7, 8⇒10, 11⇒13
15	$a = b$	∃Elim: 3, 4⇒14

7.3 Remark on Frege's Way Out

Frege in 1903 suggested weakening something like (7.7) to an assumption that may be expressed in Leśniewski's system by:

$$\forall a, b, c, d[a\varepsilon cl(c) \wedge a\varepsilon cl(d) \wedge b\varepsilon d \wedge \neg b\varepsilon cl(d) \to b\varepsilon c] \qquad (7.13)$$

It says that if one and the same thing is the class of c's and the class of d's, then extensionality holds (whatever is a d is also a c), *but it is restricted to those objects which are not the class itself*. This move is commonly referred to as *Frege's way out*.

Unfortunately, in 1938 Leśniewski showed that (7.13) and (7.9) lead to contradiction with two additional assumptions:

$$\forall a, b, c[a\varepsilon cl(c) \wedge b\varepsilon cl(c) \to a = b] \qquad (7.14)$$

$$\exists a, b, c\,[a\varepsilon a \wedge b\varepsilon b \wedge c\varepsilon c \wedge a \neq b \wedge a \neq c \wedge b \neq c] \qquad (7.15)$$

The first of them states the uniqueness of the class of objects a, the second states that there are at least three different objects:

Theorem 7.4 (Sobociński) (7.9), (7.13), (7.14), (7.15) $\vdash \perp$

The proof is somewhat tedious, so I skip it. The reader might check Sobociński's original paper for details, or (Urbaniak 2008) for a slightly simpler reconstruction.

The issue of whether Frege's way out leads to inconsistency has been debated for a while. Quine (1955) credits Leśniewski and provides his own version of the proof. Geach (1956) describes a generalization of Leśniewski's proof. Linsky and Schumm (1971) emphasize that Frege's way out, contrary to the common view, does not lead to inconsistency, but rather yields a system which is metaphysically false, severely limiting the number of existing objects (they also give an even simpler proof). They argue the system is satisfiable in a domain with only one element. Dummet (1973) argues that in fact, Frege's system is inconsistent, because he takes truth values to be objects and proves within his system that there are at least two truth values. Frege's way out is however still defended by Landini (2006). The issue is too complicated and too off-topic to be settled in this chapter. Let us get back to what Sobociński had to say about distributive classes.

7.4 Sobociński on Distributive Classes

Sobociński in his paper devoted to Leśniewski's resolution of Russell's paradox claims:

> The expression 'class(a)' in the distributive sense is nothing more than a fictitious name which replaces the well-known term of classical logic, 'the extension of the objects a'. If one takes the term in this sense, the formula '$b\varepsilon cl(a)$' means the same thing as 'b is an element of the extension of the objects a' or, more briefly, 'b is a'. In which case, the formula 'Socrates is cl(white)' means the same thing as 'Socrates is an element of the extension of white objects'; in other words: 'Socrates is white'. Thus, the understanding of 'class' in the distributive sense would reduce the formula '$b\varepsilon cl(a)$' to the purely logical formula '$b\varepsilon a$', where ' ε ' is understood as the connective of the individual proposition. (Sobociński 1949, p. 31)

Sobociński does something quite unexpected here. For some reason he reads '$\varepsilon cl(a)$' as 'is an element of the extension of the objects a'. Recall that '$cl(a)$' is just the formal version of 'the extension of the name a '. But if that is the case, ε is no longer being read by Sobociński as 'is', but rather as 'is an element of'. However, this reading is plainly wrong given the interpretation given to ε by Leśniewski.

While the reading is quite counterintuitive, it is still interesting to see what happens with the paradoxical assumptions on this reading. Replace $\alpha_1 \varepsilon cl(\alpha_2)$ with $\alpha_1 \varepsilon \alpha_2$ in those assumptions. (7.6) becomes

$$\forall a \exists b \, b\varepsilon a \qquad (7.16)$$

which is disprovable in Ontology (basically because it says that every possible name is non–empty and in Ontology one can define the empty name and obtain the negation of (7.16) by existential quantifier introduction). (7.7) is read as

$$\forall a, b, c, d[a\varepsilon c \wedge a\varepsilon d \wedge b\varepsilon d \rightarrow b\varepsilon c] \tag{7.17}$$

which, again, is disprovable in Ontology. (7.9) turns out to be provable

$$\forall b, c[b\varepsilon c \rightarrow \exists a\, a\varepsilon c] \tag{7.18}$$

(7.13) becomes:

$$\forall a, b, c, d[a\varepsilon c \wedge a\varepsilon d \wedge b\varepsilon d \wedge \neg b\varepsilon d \rightarrow b\varepsilon a] \tag{7.19}$$

which is tautologically true. (7.14) turns out to be unprovable and false in most non-trivial models:

$$\forall a, b, c[a\varepsilon c \wedge b\varepsilon c \rightarrow a = b] \tag{7.20}$$

What happens to (7.5) on this reading? It becomes:

$$\forall a, b[b\varepsilon el(a) \equiv \exists c\,(a\varepsilon c \wedge b\varepsilon c)] \tag{7.21}$$

which is certainly absurd. Take any two singular names, a, b. Clearly $a\varepsilon a \cup b \wedge b\varepsilon a \cup b$. Thus, $\exists c\,(a\varepsilon c \wedge b\varepsilon c)$. But then, $a\varepsilon el(b)$ and $b\varepsilon el(a)$, that is, it turns out that any two individuals are elements of each other. This, again, is a slightly inconvenient conclusion.

Overall, this interpretation does allow us to avoid undesired paradoxical conclusions. But this happens because quite compelling assumptions on this formalization become (at least in most of the cases) trivially false claims. This discrepancy, instead of providing a solution to the paradoxes, undermines the adequacy of Sobociński's interpretation.

7.5 Does not Ontology Contain Some Set Theory Already?

Is not our effort pointless? Is there a need to introduce cl on top of Ontology? Is not Ontology a set theory already? For instance, if one develops a standard set-theoretic semantics for Ontology (name variables range over the power set of the domain of individuals etc.), something very much like the axiom of choice can be expressed in the language of Ontology by:

$$\exists f\, \forall a, b(a\varepsilon b \rightarrow f(b)\varepsilon a) \tag{7.22}$$

which in the set-theoretic interpretation can be read as stating that there is an f such that for every non-empty set b, $f(b)$ is an element of b (see Davis 1975 for details).

Davis himself uses set-theoretic paraphrases throughout his paper, commenting briefly:

> Set theoretic paraphrases will be used throughout this investigation since they are the most
> natural to use in discussions concerning the axiom of choice (Davis 1975, p. 182).

This line of reasoning is familiar. It resembles the standard argument for the claim that the standard second-order logic is set theory in sheep's clothing.

We need to distinguish between two questions. One is whether Ontology as it is, given that it involves higher-order quantification, involves ontological commitment to sets. This issue will be tackled in Chap. 8. Another question, relevant to our current considerations, is whether it is possible to emulate within Ontology enough set theory to obtain major mathematical theories.

The mere fact that set-theoretic paraphrases are convenient in a discussion does not imply that the paraphrased language indeed says something about sets. Nor does it show that the theory in question is rich enough to go proxy for set theory. At most, one can say that *if* the language is given standard set-theoretic semantics *then* it is possible to formulate in it formulas that are true in a model iff a specific set-theoretic axiom(s) (formulated in metalanguage) hold(s) in the same model.

This might suggest that at least some facts usually stated in set theory are expressible in the language of Ontology. However, even though formulas of Ontology may have the same satisfaction conditions as certain set-theoretic claims, the language of Ontology itself does not contain a systematic device which would cmulate the behavior of the usual set-theoretic jargon.

For instance, there is a certain interesting translation along these lines:

Class a is a subset of the class b	$\forall c (c\varepsilon a \rightarrow c\varepsilon b)$
Class a is empty	$\neg \exists b\, b\varepsilon a$
Classes a and b are disjoint	$\neg \exists c\, (c\varepsilon a \wedge c\varepsilon b)$

Still, the language of Ontology is far from the generality enjoyed by the language of set theory. Let us take a closer look at (7.22). Take the common-sense set-theoretic semantics. We start with a possibly empty set of urelements, then construct the power set to obtain the range of name variables, and similarly for all other semantic categories. In this setting (7.22) says only that for any non-empty set of urelements there is a function that selects an element of that set. This is (**not**) the same as saying that for any nonempty family of pairwise disjoint nonempty sets there is a function which assigns to each element of that family a unique element of the input set. Of course, in Ontology we can emulate "nonempty family of pairwise disjoint nonempty sets of urelements" and similarly for any other "type" or "order" of the usual set-theoretic objects. But there is no single formula of Ontology which is true in a model if and only if the set-theoretic axiom of choice in its full generality holds for all semantic categories. We have to introduce an axiom-of-choice-like formula for every semantic category separately. Now, since the language of Ontology has no upper limit on how complex semantic categories can get, this means that the best we could do is to suggest a choice *schema* under which all those particular choice axioms would fall. But this is quite different from being able to express the axiom by means of a single formula of set theory.

There is another possibility to explore: perhaps, there is a sensible translation from the language of standard set theory into the language of Ontology which preserves our intuitions about distributive classes, and (hopefully) theorems of an interesting set theory. We will take a closer look at this option now.

7.6 Higher-Order Epsilon: The Basics

Recall that the question which launched the discussion about distributive classes was: how to imitate the usual talk about distributive classes in a language of Ontology or a language based on this language? As we have already seen, Leśniewski–Sobociński's solution to the Russell's paradox did not work very well: the mereological approach seems like the change of topic, and Sobociński's deflationary interpretation clearly fails.

Some preliminary work on an alternative approach has been suggested by Lejewski (1985), who tried to develop a "class-talk" within Ontology. He did so by introducing class-like operators, and a new sort of functors needed to mimic predication between such class terms. Predication operators of this sort are called higher-order epsilon operators (HOε).

Lejewski's approach was advertised in the context of Russell's paradox by (1972, pp. 42–44), who was quite optimistic. He claimed that the approach "gives one acceptable sense to the principle that there are classes of whatever sorts of objects one may specify" (Henry 1972, p. 44), and that "intuitive clarity can be maintained throughout, [a] version of the [Russell's] paradox is avoided, and *ad hoc* evasive strategems of a formalist nature shown to be needless" (Henry 1972, p. 46).

I will start with explaining briefly what higher-order epsilon operators are. Then I will consider a few ways they can be introduced. Next, I will investigate the cash value of Henry's claims and show what happens if one actually applies Lejewski's strategy to Russell's paradox.

The basic idea is that instead of treating $cl(a)$ as a name of a class, we put it in a semantic category different from the category of names and treat 'is' as ambiguous. For example, 'Socrates is a philosopher' is parsed as $s\varepsilon p$, but a sentence of a similar form, 'Reading is learning' is, if we follow Leśniewskian intuitions, not a sentence directly about objects. It rather is taken to mean that anything which reads, learns.

This sort of approach can be systematized in Leśniewski's Ontology. Define the weak identity first:

$$\forall a, b[\circ(a, b) \equiv \forall c(c\varepsilon a \equiv c\varepsilon b)] \qquad (7.23)$$

(Observe that it also holds between empty terms.) Now, every one-place sentential functor of a name argument can be nominalized:[4]

---

[4] In the original notation of the papers at issue parentheses surrounding arguments of different categories were of different shapes. For the sake of convenience I follow this aspect of the original notation, as well as I keep using ϕ as a functor variable. There is no obvious assignment of shapes of

$$\forall a, \phi[a\varepsilon trm\langle\phi\rangle \equiv a\varepsilon a \wedge \phi(a)] \tag{7.24}$$

There are at least two class-like operators we might want to introduce. One is meant to help one to speak of classes of objects satisfying a certain formula, one is meant to help one to speak of classes of objects falling under a certain term. Accordingly, read '$cl[\![\phi]\!](a)$' as 'a is the class of those objects that ϕ-ize' and '$[\![a]\!](b)$' as 'b is the class of a's'. Define these by:

$$\forall a, b[cl[\![a]\!](b) \equiv \circ(b, a)] \tag{7.25}$$

$$\forall a, \phi[cl[\![\phi]\!](a) \equiv cl[\![trm\langle\phi\rangle]\!](a)] \tag{7.26}$$

(These definitions do not interpret class terms as actually referring expressions: $cl[\![a]\!](b)$ only means that b has the same extension as a, and $cl[\![\phi]\!](a)$ only means that a refers to exactly those objects which satisfy ϕ.)

The trick now is to define a predication copula for such "fake" terms which could mimic some of our class talk. For this, we need a higher-order epsilon ε_1 defined by:

$$\forall \phi, \psi[\phi\varepsilon_1\psi \equiv \exists a\,(\phi(a) \wedge \psi(a)) \wedge \forall b, c(\phi(b) \wedge \phi(c) \rightarrow \circ(b, c))] \tag{7.27}$$

which, if you define:

$$\forall \phi[ext\langle\phi\rangle \equiv \forall b, c(\phi(b) \wedge \phi(c) \rightarrow \circ(b, c))] \tag{7.28}$$

where $ext\langle\phi\rangle$ may be read as 'ϕ' determines a unique possible name, up to coextensiveness (or, in other words, ϕ determines a class) may be expressed by:

$$\forall \phi, \psi[\phi\varepsilon_1\psi \equiv \exists a\,(\phi(a) \wedge \psi(a)) \wedge ext\langle\phi\rangle)] \tag{7.29}$$

Intuitively speaking, $\phi\varepsilon_1\psi$ is true if there is only one (up to coextensiveness) name of which ϕ can be truly predicated and it is one of the names of which ψ can be truly predicated.

The following theorem[5] indicates the contrast between the regular ε and the higher-order ε_1:

Theorem 7.5 *The following are provable in Ontology:*

$$\forall a[cl[\![a]\!]\varepsilon_1 cl[\![a]\!]] \tag{7.30}$$

$$\forall \phi[cl[\![\phi]\!]\varepsilon_1 cl[\![\phi]\!]] \tag{7.31}$$

(Footnote 4 continued)
brackets to semantic categories, I will just arbitrarily pick different shapes for different categories for the purposes of this chapter.

[5] Most of theorems and proof sketches about higher-order epsilon operators have been developed by the author.

$$\neg \forall a a \varepsilon a \tag{7.32}$$

Proof From (7.23) we get

$$\forall a[\circ(a, a) \equiv \forall c[c\varepsilon a \equiv c\varepsilon a]] \tag{7.33}$$

the right side of the equivalence follows from $\forall p(p \equiv p)$ (a trivial theorem of Protothetic). Thus $\forall a \circ (a, a)$. From (7.25) we get:

$$\forall a, b[cl[\![a]\!](a)] \tag{7.34}$$

From (7.27):

$$\forall a[cl[\![a]\!] \, \varepsilon_1 \, cl[\![a]\!] \equiv \exists b \, (cl[\![a]\!](b) \wedge cl[\![a]\!](b)) \wedge ext\langle cl[\![a]\!]\rangle)] \tag{7.35}$$

Clearly $\exists b \, (cl[\![a]\!](b) \wedge cl[\![a]\!](b))$ follows from (7.34). To show that $\forall a(ext\langle cl[\![a]\!]\rangle)$, suppose that $cl[\![a]\!](b)$ and $cl[\![a]\!](c)$. By (7.25), $\circ(a, b)$, $\circ(a, c)$, and it is trivial that $\circ$ is symmetric and transitive (that results from the symmetry and transitivity of equivalence). Thus we get (7.30). To show (7.31) just note that (7.30) implies

$$\forall \phi, a[cl[\![trm\langle \phi \rangle]\!](a)\varepsilon_1 \, cl[\![trm\langle \phi \rangle]\!](a)] \tag{7.36}$$

and use (7.26). On the other hand to obtain (7.32), define:

$$\forall a[a\varepsilon\Lambda \equiv a\varepsilon a \wedge \neg a\varepsilon a] \tag{7.37}$$

Clearly, $\neg \Lambda \varepsilon \Lambda$.

7.7 Higher-Order Epsilon and Russell's Paradox

Let us temporarily assume that we can mimic the talk about distributive classes using the above strategy. What happens if we employ it to discuss Russell's paradox as formulated by Leśniewski and Sobociński? Here is a sketch of the discussion.

Take the paradoxical axioms used earlier in this chapter. Can they be interpreted using the above strategy? It turns out that we have to play around with the language in order to be able to translate them into the language of Ontology. Also, there is no unique possible way of massaging the axioms into the language of Ontology.

- One of the options entails that only if a is a singular term there exists the class of a's, which means that this reading falls short of imitating the usual set-theoretic talk.
- Another way of translating the axioms turns (7.7) into a formula which entails that $a\varepsilon \, cl[\![b]\!]$ is equivalent to $a\varepsilon b$, which is not a plausible identification if we want

to stick with the distributive reading (without running into the already discussed conclusion that for any two individuals one is an element of another). It also seems, in this reading we lose the extensionality of distributive sets (which generally is not a good news if we want to imitate the usual language of set theory).

- Finally, I discuss yet another quite natural reading. Interestingly, translating formulas that were contradictory on the previous reading using this strategy yields a set of theorems of Ontology (which we have good reasons to believe to be consistent), but this indicates that the original set-theoretic consequence is not preserved under translation.

Now that a philosophical guide has been sketched, let us get back to the trenches.
Consider axiom 7.6:

$$\forall a \exists b \, b\varepsilon cl(a) \tag{7.38}$$

When we try to parse it using a higher-order epsilon, we see that neither $b\,\varepsilon_1\,[\![a]\!]$ nor $b\,\varepsilon\,cl[\![a]\!]$ are well-formed. The standard ε takes two expressions of category n as arguments and is of category $\frac{s}{nn}$. The ε_1 operator, on the other hand, takes two expressions of category $\frac{s}{n}$ as its arguments and is of category $\frac{s}{\frac{s}{n},\frac{s}{n}}$. There are two basic options: we can define a new higher-order epsilon of category $\frac{s}{n\frac{s}{n}}$ that mixes these two and use it in parsing the formulae, or we can somehow modify the semantic category of one of the arguments of an epsilon in the original formula. Let us start with the first option.

(i)
Take another epsilon:

$$\forall a, \phi[a\varepsilon_2\phi \equiv a\varepsilon a \wedge \phi(a)] \tag{7.39}$$

(7.6) in this reading renders:

$$\forall a \exists b \, [b\varepsilon b \wedge cl[\![a]\!](b)] \tag{7.40}$$

but this implies $\exists b \, b\varepsilon b$ which is not a theorem of Ontology. Another consequence is $\forall a \exists b \, [b\varepsilon b \wedge o(a,b)]$, and thus $\forall a[a\varepsilon a]$ (which contradicts (7.32) and is inconsistent with the axioms of Ontology).

(7.7), another axiom, after translation yields:

$$\forall a, b, c, d[a\varepsilon_2 cl[\![c]\!] \wedge a\varepsilon_2 cl[\![d]\!] \wedge b\varepsilon d \rightarrow b\varepsilon c] \tag{7.41}$$

which is equivalent to:

$$\forall a, b, c, d[a\varepsilon a \wedge cl[\![c]\!](a) \wedge cl[\![d]\!](a) \wedge b\varepsilon d \rightarrow b\varepsilon c] \tag{7.42}$$

and

$$\forall a, b, c, d[a\varepsilon a \wedge o(c,a) \wedge o(d,a) \wedge b\varepsilon d \rightarrow b\varepsilon c] \tag{7.43}$$

which is a theorem of Ontology.

When we translate another axiom, (7.9), the translation is:

$$\forall a, b[a\varepsilon b \rightarrow \exists c\, c\varepsilon_2 cl[\![b]\!]] \qquad (7.44)$$

which is equivalent to the formula:

$$\forall a, b[a\varepsilon b \rightarrow \exists c\, (c\varepsilon c \wedge cl[\![b]\!](c))] \qquad (7.45)$$

The reason why this formula should not hold is that it says that if b is non-empty, then it is coextensive with a singular c. It ultimately says that all terms are either empty or singular and thus restrict the domain cardinality to one object. (This is because we could define the universal name by $a\varepsilon V \equiv a\varepsilon a$ and apply(7.45) to it, proving that it is singular.)

The core of the problem is that (7.39), seemingly innocent, results in

$$\forall a, b[a\varepsilon_2 cl[\![b]\!] \equiv a\varepsilon a \wedge \circ(a, b)] \qquad (7.46)$$

which implies that only if b is a singular term, there is an a such that $a\varepsilon_2 cl[\![b]\!]$.

(ii)

Let us try a more subtle definition of a higher-order epsilon.[6]

$$\forall a, \phi[a\varepsilon_3 \phi \equiv a\varepsilon a \wedge \exists b\, (a\varepsilon b \wedge \phi(b))] \qquad (7.47)$$

(7.9) yields:

$$\forall a, b[a\varepsilon b \rightarrow \exists c, d\, (c\varepsilon d \wedge cl[\![b]\!](d))] \qquad (7.48)$$

which is equivalent to:

$$\forall a, b[a\varepsilon b \rightarrow \exists c, d\, (c\varepsilon d \wedge \circ(d, b))] \qquad (7.49)$$

which is valid. But let us return to (7.6) and (7.7).

When we translate (7.6) using ε_3, we get:

$$\forall a \exists b\, b\varepsilon_3 cl[\![a]\!] \qquad (7.50)$$

This is equivalent to

$$\forall a \exists b\, [b\varepsilon b \wedge \exists c\, (b\varepsilon c \wedge cl[\![a]\!](c))] \qquad (7.51)$$

i.e.

---

[6] What makes a connective a higher-order epsilon is an interesting question. The intuition is that f is an epsilon if the truth of $f(\delta, \gamma)$ requires some sort of uniqueness of the referent(s) of δ (presumably, up to coextensiveness), and some sort of inclusion between the referent(s) of δ and the referent(s) of γ. This however is far from providing a formally correct and precise definition.

$$\forall a \exists b \, [b\varepsilon b \land \exists c \, (b\varepsilon c \land \mathrm{o}(c, a))] \tag{7.52}$$

which is still not a theorem, because it implies $\exists b \, b\varepsilon b$.

(7.7) renders:

$$\forall a, b, c, d[a\varepsilon a \land \exists e \, (a\varepsilon e \land \mathrm{o}(e, c)) \land \exists e \, (a\varepsilon e \land \mathrm{o}(e, d)) \land b\varepsilon d \to b\varepsilon c] \tag{7.53}$$

which is disprovable in Ontology.

Now, 7.7 in its initial form sounds like a pretty obvious claim about distributive sets. Why does it fail in the new reading? Consider what $a\varepsilon_3 cl[\![b]\!]$ boils down to:

$$a\varepsilon a \land \exists c \, (a\varepsilon c \land \mathrm{o}(c, b)) \tag{7.54}$$

that is,

$$\exists c \, (a\varepsilon c \land \mathrm{o}(c, b)) \tag{7.55}$$

and thus it is equivalent to $a\varepsilon b$. But this identification, as we already noticed, leads to other troubles. Even more so if we try accept the corresponding translation of (7.5).

(iii)

Maybe we should go with another option and use ε_1 in our translations, but instead of name variables employ $\frac{s}{n}$-operator variables? Let us try. In this reading (7.6) yields:

$$\forall a \exists \phi \, \phi \varepsilon_1 cl[\![a]\!] \tag{7.56}$$

that is:

$$\forall a \exists \phi \, [\exists b \, (\phi(b) \land cl[\![a]\!](b)) \land ext\langle \phi \rangle] \tag{7.57}$$

which is equivalent to:

$$\forall a \exists \phi \, [\exists b \, (\phi(b) \land \mathrm{o}(a, b)) \land ext\langle \phi \rangle] \tag{7.58}$$

(7.58) follows from (7.30), which makes it a theorem.

Now, (7.7) becomes:

$$\forall \phi, a, b, c[\phi \varepsilon_1 cl[\![a]\!] \land \phi \varepsilon_1 cl[\![b]\!] \land c\varepsilon b \to c\varepsilon a] \tag{7.59}$$

that is:

$$\forall \phi, a, b, c[\exists d \, (\phi(d) \land cl[\![a]\!](d)) \land ext\langle \phi \rangle \land \exists e \, (\phi(e) \land cl[\![b]\!](e)) \land c\varepsilon b \to c\varepsilon a] \tag{7.60}$$

which is equivalent to:

$$\forall \phi, a, b, c[\exists d \, (\phi(d) \land \mathrm{o}(a, d)) \land ext\langle \phi \rangle \land \exists e \, (\phi(e) \land \mathrm{o}(b, e)) \land c\varepsilon b \to c\varepsilon a] \tag{7.61}$$

which is also a theorem.

Theorem 7.6 (7.61) *is a theorem of Ontology.*

Proof Suppose: $\exists d\,(\phi(d) \wedge \circ(a, d))$, $\forall d, e\,(\phi(d) \wedge \phi(e) \rightarrow \circ(d, e))$, $\exists e\,(\phi(e) \wedge \circ(b, e))$, $c\,\varepsilon\,b$. Thus there are some δ and γ s.t. $\phi(\delta) \wedge \circ(a, \delta)$ and $\phi(\gamma) \wedge \circ(b, \gamma)$. Also, $\phi(\delta) \wedge \phi(\gamma) \rightarrow \circ(\delta, \gamma)$. Hence: $\circ(\delta, \gamma)$. By the transitivity of $\circ$, $\circ(a, b)$ and thus $c\,\varepsilon\,a$.

The problem is that (7.6) and (7.7) originally led to contradiction. If (7.56) and (7.59) led to contradiction as well, it would have shown that Ontology is inconsistent. Fortunately an adaptation of the original proof of Theorem 7.1 does not work.

One reason is that if we use the original definition of $\star$, then while line 11 would still say $\neg\exists a\,a\varepsilon cl[\![\star]\!]$, line 12 would not contradict it. For it would say $\exists\phi\,\phi\varepsilon_1 cl[\![\star]\!]$ which reduces to:

$$\exists\phi\,[\exists b\,(\phi(b) \wedge \circ(\star, b)) \wedge \forall c, d(\phi(c) \wedge \phi(d) \rightarrow \circ(c, d))]$$

This, however is trivially the case, for $\neg\exists a\,a\varepsilon\gamma$ does an excellent job of $\phi(\gamma)$.

Perhaps, we should define $\star$ differently, employing the translation also to the class term occurring in (7.8) and using ε_1 instead of ε when talking about classes?

Strictly speaking, that cannot be the only modification, because we also need to change some of the name variable quantification into formula variable quantification. So first, define:

$$\forall a, b[vrbm\langle a\rangle(b) \equiv b\varepsilon a] \tag{7.62}$$

Then, in the original definition:

$$a\varepsilon\star \equiv a\varepsilon a \wedge \forall b(a\varepsilon cl(b) \rightarrow \neg a\varepsilon b)$$

replace '$a\varepsilon cl(b)$' with '$vrbm\langle a\rangle\varepsilon_1 cl[\![b]\!]$', thus obtaining:

$$a\varepsilon\star \equiv a\varepsilon a \wedge \forall b(vrbm\langle a\rangle\varepsilon_1 cl[\![b]\!] \rightarrow \neg a\varepsilon b)$$

There are two things to observe in this situation. One is that '$vrbm\langle a\rangle\varepsilon_1 cl[\![b]\!]$' boils down, subsequently, to:

$$\exists c\,(vrbm\langle a\rangle(c) \wedge \circ(a, c)) \wedge ext(vrbm\langle a\rangle) \tag{7.63}$$
$$\exists c\,(c\varepsilon a \wedge \circ(b, c)) \wedge \forall c, d(c\varepsilon a \wedge d\varepsilon a \rightarrow \circ(c, d)) \tag{7.64}$$
$$b\varepsilon a \wedge (a\varepsilon a \vee \neg\exists c\,c\varepsilon a) \tag{7.65}$$

Thus, the right-had side of the definition reduces to the disprovable formula:

$$a\varepsilon a \wedge \forall b[b\varepsilon a \rightarrow \neg aa\varepsilon b]$$

and the definition of $\star$ becomes just a definition of an empty term. The original derivation of the contradiction in the proof of Theorem 7.1 also fails, for line 11 ends up being $\neg\exists a\, vrbm\langle a\rangle\varepsilon_1 cl[\![\star]\!]$ and not contradicting line 12, which would say: $\exists\phi\,\phi\varepsilon_1 cl[\![\star]\!]$. The reason why there is no contradiction is that quantification over $vrmb\langle a\rangle$ terms is narrower than quantification over all formulas ϕ (for instance, $vrbm\langle a\rangle$ requires that a be non-empty, and not every formula ϕ does that).

In general

On one hand, some of the translations of problematic axioms into higher-order epsilon language clearly fail. On the other, a translation which does not fail is also not suitable, because the original consequence operation is not preserved under translation. One could ask, am I not demanding the impossible by saying that the approach fails no matter whether the translation preserves axioms or not? Not quite.

The problem with the translation which preserves all the axioms is that the system might be too weak to yield an interesting set theory.[7] The problem with translations discussed in paragraphs (i) and (ii) is not merely that the translations of the problematic axioms fail. It is that the translations fail miserably: in case of (i) by turning comprehension into something which requires the domain to be a singleton and in case of (ii) by identifying being the class of b's with being a b. An interesting translation would most likely loose some axioms, but would give us some sensible grounds to do that and would not lead to ridiculous conclusions. Perhaps, there are such translations into the language of Ontology, but none of the known higher-order epsilon approaches provides one.

Considerations of this section provide a good argument against Henry's optimistic claim about the importance of the higher order epsilon. At least for now the view that set theory can be interestingly imitated within the language of Ontology is highly implausible.

7.8 Słupecki's Generalized Mereology

Słupecki (1958)[8] attempted to obtain a sensible set theory within Leśnewskian framework by extending Leśniewski's mereology. As he put it:

> Generalizing mereology, I keep in mind the following two aims:
>
> (a) in the generalized system the notion of a set should preserve its intuitive sense; in particular, sets should be defined in a way that treats them as concrete objects.

---

[7] To be fair, details and weaknesses of the translations discussed in paragraph (iii) require further investigation, which is beyond the scope of this book.

[8] The problem with reconstructing Słupecki's generalized mereology is that the only paper that he published which pertained to it is a four-page abstract, so quite a few details have to be filled out by the reader. What follows is my best shot and understanding what the system looks like.

(b) the generalized system of mereology should be able to underlie mathematics to the same extent to which simple type theory or axiomatic set theory does. This purpose has not been attained by Leśniewski's own system, in which, for instance, natural number arithmetic cannot be developed (Słupecki 1958, p. 156).[9]

Słupecki, instead of Ontology (with name variables $a, b, c, \ldots$ uses a language with individual variables $x, y, z, \ldots$ predicate variables $f, g, f_1, g_1, \ldots$ and general functor variables $\phi, \phi_1, \ldots$ (whose category is to be contextually salient), and he often uses formulas with free variables as if the free variables occurring in them were bound by a universal quantifier standing in the front of the formula. His mereological axioms mimic Leśniewski's axioms for Mereology. He assumes the reflexivity and transitivity of (improper) parthood relation and assumes extensionality, pretty much as Leśniewski's axioms (a–c) of the 1920 axiomatization of Mereology.[10] Słupecki's axiom 4 states the existence and uniqueness of a fusion of objects satisfying a predicate, and it follows from axiom (d) (which states the existence of the fusion) and other axioms of the 1920 formulation of Mereology.[11] Słupecki also provides certain assumption that go beyond what Leśniewski has done. His axiom 5 postulates a certain fixed object to go proxy for the empty set,[12] and he defines a few interesting notions that are supposed to allow us to mimic the language of set theory.

Słupecki starts with his version of Mereology in a fairly standard higher-order language. He introduces an additional constant $\leq$ which stands for parthood relation. Axioms that he introduced more or less mirror Leśniewski's axioms:

$$\textbf{A.1.} \quad x \leq x$$

$$\textbf{A.2.} \quad x \leq y \wedge y \leq z \rightarrow x \leq z$$

$$\textbf{A.3.} \quad x \leq y \wedge y \leq x \rightarrow x = y$$

Definitions of overlapping and fusion are fairly standard:

$$\textbf{D.1.} \quad x \dagger y \equiv \exists z\, (z \leq x \wedge z \leq y)$$

$$\textbf{D.2.} \quad x \delta f \equiv \forall z [f(z) \rightarrow z \leq x] \wedge \forall y \{y \leq x \rightarrow \exists z\, [f(z) \wedge z \dagger y]\}$$

---

[9] "Uogólniając mereologię kieruję się następującymi dwoma celami: (a) w systemie uogólnionym pojęcie zbioru powinno zachować niezmieniony sens intuicyjny, w szczególności zbiory powinny być tak zdefiniowane, by były konkretnymi przedmiotami; (b) uogólniony system mereologii powinien nadawać się do podbudowania matematyki w tym zakresie, w jakim służą do tego celu prosta teoria typów logicznych lub aksjomatyczna teoria mnogości. Celu tego nie spełnia pierwotny system Leśniewskiego, w którym nie może być np. zbudowała arytmetyka liczb naturalnych."

[10] The background logic is weaker but the properties of the parthood relation are essentially the same.

[11] Basically, the uniqueness follows from extensionality.

[12] So Słupecki's system, unlike Leśniewski's, is not free of existential assumptions.

The category of f is $\frac{s}{n}$, it is just a predicate of a first order language and we read '$x\delta f$' as 'x fuses f'. The last axiom states the uniqueness of fusions of nonempty predicates:

$$\textbf{A.4.} \quad \exists y\, f(y) \rightarrow \exists! x\, x\delta f$$

In the language of Leśniewski's Mereology the above formulas would have the following form:

$$\forall a(a\varepsilon a \rightarrow a\varepsilon ingr(a)) \tag{7.66}$$

$$\forall a, b, c(a\varepsilon ingr(b) \wedge b\varepsilon ingr(c) \rightarrow a\varepsilon ingr(c)) \tag{7.67}$$

$$\forall a, b(a\varepsilon ingr(b) \wedge b\varepsilon ingr(a) \rightarrow (a\varepsilon b \wedge b\varepsilon a)) \tag{7.68}$$

$$\forall a, b(a\varepsilon orl(b) \equiv \exists c\, (c\varepsilon ingr(a) \wedge c\varepsilon ingr(b))) \tag{7.69}$$

$$\forall a, b(a\varepsilon fus(b) \equiv a\varepsilon a \wedge \forall c(c\varepsilon b \rightarrow c \leq a) \wedge \forall c(c \leq a \rightarrow \exists d\, (d\varepsilon b \wedge d\varepsilon orl(c)))) \tag{7.70}$$

$$\forall b(\exists a\, a\varepsilon b \rightarrow \exists c\, (c\varepsilon fus(b) \wedge \forall d(d\varepsilon fus(b) \rightarrow d\varepsilon c))) \tag{7.71}$$

From now on I will translate Słupecki's formulas into the language of Leśniewski's mereology without giving their original form first.

The first thing that Słupecki wants to account for is the fact that when we speak of classes in natural language we do seem to refer to the empty set. He introduces a name constant 0, which he also axiomatically assumes to refer to an object. Thus, he accepts as an axiom:

$$\exists a\, 0\varepsilon a \tag{7.72}$$

where '0' is supposed to be "a name of an arbitrary but fixed object" (p. 156). Then, in his generalized mereology the definition of a set has the following form (for the "Słupecki class" I use the symbol CL_S, to differentiate it from the Leśniewski's notion of class):

$$\forall a, b(a\varepsilon CL_S(b) \equiv (a\varepsilon fus(b) \wedge \exists c\, c\varepsilon b) \vee (\neg\exists c\, c\varepsilon b \wedge a\varepsilon 0)) \tag{7.73}$$

The basic idea here is that the class of b's is the fusion of b's if the name is non–empty and it is identified with the selected object 0 otherwise. He also indicates how classes generated by expressions of various categories can be defined. The class of objects that correspond to an expression b of category n is just $CL_S(b)$. The class generated by an expression ϕ of category $\frac{s}{n}$ is defined by:

$$\forall a, \phi(a\varepsilon\Delta_\phi \equiv a\varepsilon a \wedge \exists b\, (\phi(b) \wedge a\varepsilon Cl_S(b))) \tag{7.74}$$

Intuitively, for a to be the class of ϕ, there has to be a possible name b such that $\phi(b)$ holds and a fuses b. 'Δ_ϕ' names all and only those fusions of b's for which it is true that $\phi(b)$. For expressions of category $\frac{s}{\frac{s}{n}}$ the definition is quite similar:

$$\forall a, \Phi(a\varepsilon\Delta_\Phi = a\varepsilon a \land \exists\phi\,(\Phi(\phi) \land a\varepsilon Cls(\Delta_\phi))) \tag{7.75}$$

That is, take an $\frac{s}{n}$-formula ϕ. If $\Phi(\phi)$, then all things which are $Cls(\Delta_\phi)$'s are also among the Δ_Φ's.

Observe that both Δ_ϕ and Δ_Φ can be plural terms and refer to multiple objects. For instance, if 'ϕ' stands for 'is a human being', then Rafal $\varepsilon\Delta_\phi$ because Rafal ε Rafal and Rafal ε human being and Rafal ε Cls(Rafal). The same holds for any particular proper name of a human being, and thus each particular human is denoted by '$\Delta_{\text{is a human being}}$'.

To obtain a single class, one more application of the Cls operation is needed: $Cls(\Delta_\phi)$ is the class corresponding to ϕ and $Cls(\Delta_\Phi)$ is the class corresponding to Φ. (So, for instance, to obtain the class of $\phi=$'denotes exactly two lizards', you first take fusions of names which denote exactly two lizards, and then fuse those fusions into one class.) Definitions of classes corresponding to expressions of higher categories are analogous.

Another notion which Słupecki thinks needs redefining is the notion of being an element of a set. This is because Leśniewski's definition of el eventually identifies being an element of a set with being its ingredient. In order to introduce a more fine-grained notion, Słupecki defines three auxiliary notions.

$$\forall a, b, c(\delta_1(a, b, c) \equiv a\varepsilon c \land a\varepsilon ingr(b)) \tag{7.76}$$

$$\forall a, b, c, \phi(\delta_2(a, b, c, \phi) \equiv \phi(c) \land a\varepsilon ingr(b)) \tag{7.77}$$

$$\forall a, b, \phi, \Phi(\delta_3(a, b, \phi, \Phi) \equiv \Phi(\phi) \land a\varepsilon ingr(b)) \tag{7.78}$$

To explain the notion of being an element Słupecki introduces the following abbreviations (mind that they are not definitions because they explain the meaning of $\in_1, \in_2, \in_3, \ldots$ only in certain contexts).

$$\forall a, b(a \in_1 Cls(b) \equiv \delta_1(a, Cls(b), b)) \tag{7.79}$$

$$\forall a, \phi(Cls(a) \in_2 Cls(\Delta_\phi) \equiv \delta_2(Cls(a), Cls(\Delta_\phi), a, \phi)) \tag{7.80}$$

$$\forall\phi, \Phi(Cls(\Delta_\phi) \in_3 Cls(\Delta_\Phi) \equiv \delta_3(Cls(\Delta_\phi), Cls(\Delta_\Phi), \phi, \Phi)) \tag{7.81}$$

So, being an element of $Cls(b)$ an object has to be b and part of $Cls(b)$, for $Cls(a)$ to be an element of $Cls(\Delta_\phi)$, the former has to be a part of the latter and $\phi(a)$ has to hold, and so on...Basically apart form being a part of a class, Słupecki added an intensional element of satisfying a certain condition used in the construction of this class.

Now, Słupecki in order to obtain a "copy" of the language of set theory invites us to take all theorems of his mereology and throw out all the theorems which are not constructed from:

$$a \in_1 Cl_S(b), Cl_S(a) \in_2 Cl_S(\phi), Cl_S(\Delta_\phi) \in Cl_S(\Delta_\Phi), \ldots \quad (7.82)$$

together with propositional connectives and quantifiers. The intersection of the set of theorems of his mereology with this language is dubbed 'system **M** of generalized mereology'.[13]

Now, things get a bit more hasty in Słupecki. From among theorems of Ontology he distinguishes theorems constructed from:

$$a\varepsilon b, \phi(a), \Phi(\phi), \ldots \quad (7.83)$$

together with propositional connectives and quantifiers. The system thus separated is called S_1.

Now, he claims that there is a a mapping from the language of **M** onto the language of S_1. Every formula of the form of the n-th element of the sequence in (7.83) is assigned to a formula of the form of the n-th element of the sequence in (7.82). So, $a \in_1 CL_S(b)$ is mapped to $a\varepsilon b$, $CL_S(a) \in_2 CL_S(\phi)$ is mapped to $\phi(a)$, and so on. If α_1 and β_1 are assigned to α and β, f_1 is a one-place sentential connective, f_2 is a two-place sentential connective, then assign: $f_1\alpha_1$ to $f_1\alpha$, $\alpha_1 f_2\beta_2$ to $\alpha f_2\beta$, $\exists\tau\,\alpha_1$ to $\exists\tau\,\alpha$ and $\forall\tau\alpha_1$ to $\forall\tau\alpha$.

The following are theorems of Słupecki's mereology (before the language is restricted):

$$\forall a, b(a \in_1 Cl_S(b) \equiv a\varepsilon b) \quad (7.84)$$

$$\forall a, \phi(Cl_S(a) \in_2 Cl_S(\Delta_\phi) \equiv \phi(a)) \quad (7.85)$$

$$\forall \phi, \Phi(Cl_S(\Delta_\phi) \in_3 Cl_S(\Delta_\Phi) \equiv \Phi(\phi)) \quad (7.86)$$

and so on. Hence the mapping between S_1 and **M** establishes an isomorphism between the theories. Since S_1 in quite an obvious manner corresponds to the language of simple type theory, Słupecki's comment is:

> It follows that system **M**, and even more the whole system of generalized mereology suffices as a system of foundations of mathematics, because—as it is well-known—S_1 suffices for that purpose. This conclusion may seem trivial because the full system of simple type theory without any additions can constitute a foundation of mathematics. However, in my paper I give arguments (I realize that those arguments are not wholly convincing and for that reason I do not repeat them here) indicating that constants belonging to semantics categories to

---

[13] It is a fairly awkward procedure, for the axioms of the theory no longer are theorems of system **M**, but the set of theorems of **M** is nevertheless well defined. Perhaps, there is an axiomatization of **M** in the language of **M**. This, however, lies beyond the scope of present considerations (in particular, because I think that as a replacement for set theory **M** is flawed anyway). This does not mean that the system is not interesting in itself.

which variables $f, \phi, \Phi, \ldots$ belong are not names of sets or any other objects, but rather expressions devoid of independent reference. Those terms cannot denote concrete objects, so the full system of simple type theory does not fulfill aim (a), which was mentioned above (p. 158).[14]

7.9 Challenges to Słupecki's Account

Słupecki's proposal meets several challenges. Notice first that introducing (7.72)—an arbitrary object to go proxy for the empty set—is a rather unintuitive technical trick. The theory ceases to be ontologically neutral. Also, philosophically speaking, there seems to be no particular object which we could pick out and determine to do the job of the empty set. On Leśniewski's view, technical niceties that this move provides do not constitute a reason for departing from our intuitions about the empty set. And one of the main intuitions is that if we think of sets mereologically there is no reason to say that there is an empty set (and if we do not think of sets mereologically, it seems clear that the empty set is not just an arbitrarily picked and fixed concrete object).

But there is a more severe problem with Lsłupecki's account of the empty set. Consider constant '0' as introduced by axiom (7.72). The following is a theorem of Ontology:

$$\forall a[0\varepsilon a \rightarrow 0\varepsilon fus(0) \wedge \exists c\, c\varepsilon 0]$$

(0 itself being a witness for the existential claim). This means, by definition 7.73, that $0\varepsilon CL_S(0)$. Also, it follows that $0\varepsilon 0 \wedge 0\varepsilon ingr(CL_S(0))$ and thus:

$$\delta_1(0, CL_S(0), 0)$$

This, by definition (7.79) means $0 \in_1 CL_S(0)$. But given that $CL_S(0)$ simply *is* 0, this means $0 \in_1 0$, which is a disturbing consequence given that 0 was supposed to be empty.

Another problem stems from the fact that Słupecki defines classes in mereological terms and elementhood in terms of satisfying certain conditions. This leads to situations in which one and the same class might be determined by non-equivalent conditions and thus have different members. Let us see an example of this. Define:

---

[14] "Stąd też wynika, że system **M**, a tym bardziej cały system uogólnionej mereologii, wystarcza do ugruntowania matematyki, gdyż—jak wiadomo—do celu tego wystarcza system S_1. Wniosek ten może wydać się trywialny z tego względu, że pełny system prostej teorii typów może być bez żadnych uzupełnień podbudową matematyki. W pracy podaję jednak argument (zdaję sobie sprawę, że argumenty te nie są całkowicie przekonywające i dlatego tu ich nie powtarzam) przemawiający za tym, że terminy stałe, należące odpowiednio do tych kategorii semantycznych do których należą zmienne $f, \phi, \Phi, \ldots$ nie są nazwami zbiorów ani też żadnych innych przedmiotów, lecz wyrażeniami pozbawionymi samodzielnego sensu. Terminy te nie mogą też oznaczać przedmiotów konkretnych, a więc pełny system prostej teorii typów nie spełnia celu (a), o którym poprzednio była mowa."

$$\forall a \, (\phi'(a) \equiv a \, \varepsilon \, a \land (a \, \varepsilon \, a \lor \neg a \, \varepsilon \, a)) \tag{7.87}$$

$$\forall a \, (\phi''(a) \equiv (a \, \varepsilon \, a \lor \neg a \, \varepsilon \, a)) \tag{7.88}$$

The claim is:

Theorem 7.7 $Cl_S(\Delta_{\phi'})$ *is the same object as* $Cl_S(\Delta_{\phi''})$.

Proof Indeed:

$$\forall a[a \varepsilon \Delta_{\phi'} \equiv (a \varepsilon a \land \exists b \, (\phi'(b) \land a \varepsilon Cl_S(b)))]$$

That is,

$$\forall a[a \varepsilon \Delta_{\phi'} \equiv (a \varepsilon a \land \exists b \, ((b \varepsilon b \land (b \varepsilon b \lor \neg b \varepsilon b) \land a \varepsilon Cl_S(b))))]$$

$$\forall a[a \varepsilon \Delta_{\phi'} \equiv (a \varepsilon a \land \exists b \, (b \varepsilon b \land a \varepsilon Cl_S(b)))]$$

But if $a \varepsilon a$ then also $a \varepsilon Cl_S(a)$. Hence:

$$\forall a[a \varepsilon \Delta_{\phi'} \equiv a \varepsilon a] \tag{7.89}$$

Also,

$$\forall a[a \varepsilon \Delta_{\phi''} \equiv (a \varepsilon a \land \exists b \, (\phi''(b) \land a \varepsilon Cl_S(b)))]]$$

$$\forall a[a \varepsilon \Delta_{\phi''} \equiv (a \varepsilon a \land \exists b \, ((b \varepsilon b \lor \neg b \varepsilon b) \land a \varepsilon Cl_S(b)))]]$$

$$\forall a[a \varepsilon \Delta_{\phi''} \equiv (a \varepsilon a \land \exists b \, (a \varepsilon Cl_S(b)))]]$$

But since if $a \varepsilon a$ then $a \varepsilon Cl_S(a)$, it also follows that if $a \varepsilon a$ then $\exists b \, (a \varepsilon Cl_S(b))$. Therefore:

$$\forall a[a \varepsilon \Delta_{\phi''} \equiv a \varepsilon a] \tag{7.90}$$

But this is problematic. (7.85) together with (7.87) imply that:

$$\forall a(Cl_S(a) \in_2 Cl_S(\Delta_{\phi'}) \equiv \phi'(a) \equiv a \varepsilon a)$$

and together with (7.88) it implies that:

$$\forall a(Cl_s(a) \in_2 Cl_S(\Delta_{\phi''}) \equiv \phi''(a) \equiv a \varepsilon a \lor \neg a \varepsilon a)$$

It seems that at least certain instances of the indiscernibility of identicals have to be rejected if Słupecki's theory is to be maintained. In this case, we should reject:

$$\forall a, \phi, \psi [Cl_S(\Delta_\phi) = Cl_S(\Delta_\psi) \rightarrow (Cl_S(a) \in_2 Cl_S(\Delta_\phi) \equiv Cl_S(a) \in_2 Cl_S(\Delta_\psi))]$$
$$(7.91)$$

Otherwise, the theory would imply that $\forall a a \varepsilon a$, which contradicts an already provable theorem of Ontology.

Therefore it seems Słupecki's generalized mereology has certain flaws that preclude it from being a good candidate for a nominalist replacement for the language of any mainstream (and extensional) set theory. Even if the theory was devoid of oddities of the kind mentioned above, it still would not be obvious that it follows the Leśniewskian agenda and satisfies Słupecki's requirements.

It is also not clear whether isomorphism with type theory is enough. When motivating his work, Słupecki insists that Ontology is not sufficient because there is no arithmetic within it (presumably because the axiom of infinity is missing, but Słupecki does not specify). After a while, Słupecki claims his system does the job because it is isomorphic with a subsystem of Ontology. How can one reasonably hold both claims, is not clear. Also, if a subsystem of Ontology is good enough, it is not clear why anything else is needed on top of it.

Suppose Słupecki has a way around these philosophical qualms. Another worry remains. Słupecki insisted that the system should be nominalistically acceptable, but he uses higher-order quantification in his system, and whether this device is nominalistically acceptable is far from obvious. In fact, the next chapter is devoted to this question.

References

Davis, C. (1975). An investigation concerning the Hilbert-Sierpiński logical form of the axiom of choice. *Notre Dame Journal of Formal Logic, 16*, 145–184.

Dummet, M. (1973). Frege's way out: A footnote to a footnote. *Analysis, 33*, 139–140.

Geach, P. (1956). On Frege's way out. *Mind, 65*, 408–408.

Henry, D. (1972). *Medieval logic and metaphysics: A modern introduction.* London: Hutchinson.

Landini, G. (2006). The ins and outs of Frege's way out. *Philosophia Mathematica, 14*, 1–25.

Lejewski, C. (1985). Accommodating the informal notion of class within the framework of Leśniewski's Ontology. *Dialectica, 39*, 217–241.

Linsky, L., & Schumm, G. (1971). Frege's way out: A footnote. *Analysis, 32*, 5–7.

Quine, W. V. (1955). On Frege's way out. *Mind, 64*, 145–159.

Słupecki, J. (1958). Towards a generalized Mereology of Leśniewski. *Studia Logica, 8*, 131–154.

Sobociński, B. (1949). L'analyse de l'antinomie Russellienne par Leśniewski. *Methodos, 1–2*(1, 2, 3; 6–7):94–107, 220–228, 308–316, 237–257 [translated as "Leśniewski's analysis of Russell's paradox" (srzednicki and Ricky, 1984, 11–44)].

Urbaniak, R. (2008). Leśniewski and Russell's paradox: Some problems. *History and Philosophy of Logic, 29*(2), 115–146.

Chapter 8
Nominalism and Higher-Order Quantification

Abstract Leśniewski's main motivations were nominalistic. Yet, it is unclear whether quantifiers in his systems commit one to the existence of sets, because he described them only proof-theoretically and they seem to behave like higher-order quantifiers. I first (tentatively) argue that Ontology is committed to sets. The key assumption in the argument is that the logic obtained with the substitutional semantics lacks certain desired features which the logic with set-theoretic semantics has. I describe two known attempts to provide Ontology with a nominalistically acceptable semantics (Simons, Rickey). I find both these attempts lacking, even though the basic intuitions behind them are quite compelling. I employ those intuitions to develop a more satisfactory semantics, show how the tentative critical argument fails from this perspective and defend the philosophical viability of this approach.

8.1 Introductory Remarks

Leśniewski was a nominalist and he developed Prototethic, Ontology and Mereology in his attempt to provide nominalistically acceptable foundations of mathematics. Yet, a philosophical question arises, even prior to the question whether the systems can emulate interesting mathematical theories. Is the logical part of the theory (Prototethic and Ontology) ontologically innocent and nominalistically acceptable?

In particular, it is not clear whether a nominalist can use Leśniewski's quantification, especially in Ontology. What, for instance, do name variables in Ontology range over? Leśniewski himself did not provide any intuitive explanation of his quantifiers and the first stab might be that it is subsets of the domain of individuals that have to be values of bound variables of this sort. If this really is the case, anyone who uses the system will fail in providing a nominalistic reconstruction of mathematics.

I'll start with a few remarks about the notion of ontological commitment. Then I'll introduce a system dubbed *Quantified naming logic*. I will then present an argument to the effect that this system requires set-theoretic semantics and is

R. Urbaniak, *Leśniewski's Systems of Logic and Foundations of Mathematics*,
Trends in Logic 37, DOI: 10.1007/978-3-319-00482-2_8,
© Springer International Publishing Switzerland 2014

therefore committed to the existence of sets. Next, I will discuss a potential response to this worry,which instead of set-theoretic semantics employs substitutional approach to higher-order quantifiers. The main problem with this approach seems to be that a nominalist will run out of substitution instances to emulate standard quantification over an infinite domain. As another alternative to set-theoretic semantics, we'll look at Simon's combinatorial semantics. While philosophically compelling, it is technically underdeveloped and for this reason cannot constitute the whole story. Then, I will look at yet another approach, Rickey's natural models and point out its weak spots. Finally, I will try to make a small step forward and introduce relational semantics for plural quantification, arguing for its philosophical plausibility. I will end with a brief list of issues that need to be addressed.

Before we turn to the question whether we are ontologically committed to sets when we use a certain type of quantification, a few words of explanation regarding the notion of ontological commitment (of logical theories) are due. I take (interpreted) theories to be the primary bearers of ontological commitment. People become ontologically committed when they accept ontologically committed theories.

Quine's way of reading off the ontological commitment of a theory, in the stronger reading, consists in the conjunction of two claims: (i) each such theory should be first translated into a regimented, first-order and extensional language, and (ii) the initially considered theory is committed to the existence of those things that have to belong to the range of first-order quantifiers of the regimented version thus obtained.[1]

I tend to disagree with Quine.[2] Even putting (i) above aside, I do not think that there is a "criterion" of ontological commitment that both analyzes our intuitions about ontological commitment correctly and is epistemically more tractable than the notion of ontological commitment itself. It does not seem that our understanding of the relation 'x has to be in the range of the first-order version of a theory T for T to be true' is any clearer that our notion of ontological commitment itself. One of the worries is, for instance, that many theories formulated in natural language, even if they have a first-order version,[3] do not have a unique first-order version, and those non-unique versions differ on what has to lie in their range of quantification in order for them to be true.[4]

Even though we cannot always read off our ontological commitment from the way we ordinarily talk (if I do something for the sake of convenience, am I really committed to an object which is the sake of convenience?), when we proceed with a first-orderization of our theories our regimentation can be guided by some prior metaphysical conceptions that we are inclined to end up with anyway: and depending on what restrictions we put on our procedure we might end up with different ontologies produced by the commitment of our first-order quantifiers. Agreement

---

[1] The weaker reading requires only that the language be first-order, without requiring that it be the extensional language familiar from first-order classical logic.

[2] I mostly agree with the criticism of this sort of approach to be found in Chihara (1973, 1990).

[3] The notion of being a first-order version is also not very clear, but let us not worry about that and just assume that what matters is agreement on theorems *modulo* some fixed translation.

[4] These and related issues lie beyond the scope of my present considerations, though.

on theorems (*modulo* translation) is not enough to ensure the uniqueness of our underlying ontology.[5]

This, however, does not mean that I should not say a few words to elucidate what it means for a *logic* to carry ontological commitment to sets. I take it to mean that even if someone's prior overall beliefs did not commit one to sets, one becomes committed to sets when one accepts a given logic as a reasonable tool for one's practical and theoretical purposes. For now, let me adopt the following rough–and–ready requirement: a logic is committed to sets if (not necessarily iff) its (preferred) semantics[6] requires that certain variables of its language range over sets (or that certain constants of its language name sets) for the theorems of this logic to be valid. If an equally successful explanation can be given which does not require the existence of sets I will say that the logic (with this new semantics) does not commit one to sets.

8.2 Quantified Naming Logic

In this chapter I will be concerned only with a subsystem of Ontology (it can be thought of as a variant of the logic of plurals), dubbed Quantified Name Logic (QNL). One of the reasons why plural quantification is worth bringing up in this context is that by the time the logic of plurals came to attention, a debate about the ontological commitment of plural quantification has already taken place, dressed up as a debate about interpreting Leśniewskian quantifiers. The logic in question will be the consequence operation that the language of QNL yields given set-theoretic semantics (details will follow).[7]

In QNL, name variables are the only admissible kind of variables and the only quantifiers are those that bind name variables. Well-formed formulas are constructed from name variables $a_1, a_2, a_3, \ldots \in Var$, Boolean connectives $\neg, \wedge$, a sentential functor of two name arguments ε (read as 'is one of' or simply 'is'), and the existential quantifier $\exists$ according to the following rules:

(i) If $\alpha_1, \alpha_2 \in Var$, then $\alpha_1 \, \varepsilon \, \alpha_2$ is a well-formed formula.
(ii) If ϕ_1, ϕ_2 are wff's and $\alpha \in Var$, then $\neg(\phi_1)$, $(\phi_1) \wedge (\phi_2)$ and $\exists \alpha \, (\phi_1)$ are wff's.
(iii) Nothing else is a wff.

Recall that intuitively name variables behave like place-holders for countable name phrases (henceforth called *names*), no matter whether those are empty, singular or refer to multiple objects. Under an interpretation '$a \, \varepsilon \, b$' is true iff a "names" exactly one object and this object is also "named" by b (b may "name" other objects as well, but it does not have to). Some examples of QNL renderings of natural language sentences are:

---

[5] This point shines also through some of Smith's remarks (2009) about Parsons (2008).

[6] And I take a logic to be given together with a semantics, more remarks on this issue will follow.

[7] Sometimes, I will use 'QNL' and 'the logic of plurals' interchangeably, trusting this will not cause any confusion.

Socrates is a philosopher	Socrates ε philosophers
All cats are animals	$\forall a\,(a\,\varepsilon\,\text{cats} \to a\,\varepsilon\,\text{animals})$
Some logicians admire only each other	$\exists a\,[\forall b\,(b\,\varepsilon\,a \to b\,\varepsilon\,\text{logicians})\wedge$
	$\wedge\forall c\,\forall d\,(c\,\varepsilon\,a \wedge c\,\text{admires}\,d\wedge$
	$\wedge d\,\varepsilon\,d \to \neg d\,\varepsilon\,c \wedge d\,\varepsilon\,a)]$

As a nominalist, Leśniewski intended to develop his systems in a nominalistically acceptable way. This is why, for instance, he meticulously built his inscriptional syntax to ensure it does not bear commitment to expression types. Alas, he did not elaborate on the semantics of Ontology. He simply treated quantification in Ontology as primitive.

As the notion of formal semantics firmly established its position in the minds of mid-twentieth century logicians, this became philosophically suspicious. A question arose as to how we are to make sense (nominalistically) of Leśniewski's quantification.[8] Before looking at various attempts of dealing with this issue, let us take a look at the reason why one might find higher-order quantification nominalistically problematic.

Main semantics for plural quantification are either set-theoretic or substitutional. One of the main argument for the claim that plural quantification commits one to sets relies on rejecting the latter. Once one is left with the former, sets are in the range of quantification, and thus one is committed to them when using the system. To look at this argument, let's review these semantics briefly.

A **set-theoretic QNL-model** is a structure $\langle D, I_{set}\rangle$ where D is a non-empty domain of objects and I_{set} maps Var into the powerset of D. A **substitutional QNL-model** is a structure $\langle N, I_{sub}, Val\rangle$ where $N \neq \emptyset$ is a set of name substituends, I_{sub} maps Var into N, and Val maps the set of pairs $\{\langle x, y\rangle \mid x, y \in N\}$ into $\{1, 0\}$.

Under the substitutional interpretation the truth of substitution instances of atomic formulas is taken to be primitive and Val is a function that assigns truth values to such instances. Pure substitutional interpretation explicitly refuses to provide further analysis of truth-conditions of substitution instances of atomic formulas, especially in terms of the reference of their constituents. As for satisfaction, I skip the clauses for Boolean connectives and describe how it works for atomic and quantified formulas:

$$\langle D, I_{set}\rangle \models_{set} \alpha\,\varepsilon\,\beta \text{ iff } \mid I_{set}(\alpha) \mid= 1, \text{ and } I_{set}(\alpha) \subseteq I_{set}(\beta).$$

$$\langle D, I_{set}\rangle \models_{set} \exists\alpha\,\phi \text{ iff } \langle D, I_{set}^{\alpha}\rangle \models_{set} \phi,$$

for some I_{set}^{α} which differs from I_{set} at most at α.

$$\langle N, I_{sub}, Val\rangle \models_{sub} \alpha\,\varepsilon\,\beta \text{ iff } Val(I_{sub}(\alpha), I_{sub}(\beta)) = 1.$$

$$\langle N, I_{sub}, Val\rangle \models_{sub} \exists\alpha\,\phi \text{ iff } \langle N, I_{sub}^{\alpha}, Val\rangle \models_{sub} \phi,$$

for some I_{sub}^{α} which differs from I_{sub} at most at α.

---

[8] For debate concerning the interpretation of Leśniewski's variables see for instance (Lejewski 1954b; Prior 1965; Küng and Canty 1970; Sagal 1973; Küng 1974, 1977; Kielkopf 1977; Simons 1985, 1995; Rickey 1985). See Appendix II for more details.

That is, set-theoretic semantics takes name variables to range over subsets of the domain, whereas the substitutional semantics takes an existentially quantified statement to be satisfied if one of its instances (with respect to the variable involved) is satisfied.[9, 10]

8.3 The Received View

What I call the received view (motivated by certain remarks by Quine, Gödel or Skolem) is the view that any logic that uses higher-order quantification comparable to at least monadic second-order logic is committed to the existence of sets. One of the reasons why one might think that plural quantification bears commitment to abstract objects is the belief that plural quantifiers have to range over sets, possibly because another alternative, the substitutional reading of plural quantifiers is not satisfactory. One could imagine someone arguing as follows.

[SA1] Whenever we present a logical system (involving quantification) we have to provide it with a formal semantics which we think captures the meaning of these quantifiers adequately.

[SA2] If we use a certain logical system and believe that a certain semantics is an adequate semantics of this language, we are committed to whatever lies in the range(s) of quantification of some type of variables of the object language according to this semantics.

---

[9] It may seem that additional requirements should be put on Val so that certain formulas come out valid. For instance $\forall a, b, c\, (a\,\varepsilon\,b \wedge b\,\varepsilon\,c \rightarrow a\,\varepsilon\,c)$ is set-theoretically valid, but not substitutionally valid. (Take the interpretation where $N = \{1, 2, 3\}$, $Val(1, 2) = Val(2, 3) = 1$ but $Val(1, 3) = 0$, and $I(a) = 1, I(b) = 2, I(c) = 3$.) Actually, the point of the substitutional interpretation was to allow for disagreements of this sort (Dunn and Belnap 1968). For our present concerns, these issues aren't important.

[10] It is worth remarking that providing QNL with a set-theoretic semantics (arguably) deprives them of the expressive power that it has been argued the natural language plural quantification has. A classical argument is this. The following is intuitively true:

(8.1) There are some objects such that all and only those objects that are sets and not elements of themselves are among them.

The following, however, cannot be the case (the sentence states the existence of Russell's set):

(8.2) There is a set of objects such that all and only those objects that are sets and not elements of themselves are elements of it.

(8.1) seems to state the existence of a certain plurality (the plurality of those objects that are sets and not elements of themselves). So, *prima facie*, there are pluralities which are not sets. Hence, quantification over pluralities cannot be fully captured by interpretation which reads it as quantification over sets (or subsets of a domain).

For now, without any attempt to explain the use of plurals which does not seem to be captured by the set-theoretic semantics I will just remark that the expressivity provided by the set-theoretic semantics is complex enough to raise problems that will be my main concern.

[SA3] The logic of plurals can have two kinds of semantics: a set-theoretic seman-
       tics (discussed before) or a substitutional semantics.[11]

[SA4] QNL with nominalistically acceptable substitutional semantics is not the-
       oretically satisfactory.

[SA5] Therefore it has to be given a set-theoretic semantics.

[SA6] But on the set-theoretic reading of plural quantifiers, they range over subsets
       of a domain.

[SA7] Subsets of a domain are abstract objects and hence whenever we use the
       logic of plurals we are committed to abstract objects.

Let us dub arguments that go along these lines **semantic arguments** for the claim
that higher-order quantification commits one to abstract objects.[12]

---

[11] For instance(Kearns 1969,165–166) says:

> I feel that one can distinguish two fundamentally different ways of regarding variables– I will
> call these two views of variables. The first view I call the Russell–Quine view; the second is
> the Frege–Leśniewski view (these will be abbreviated as R–Q and F–L, respectively). These
> two are not the only possible views, but I feel that they are the two basic views; other views
> will be variants of one or the other, or perhaps combinations of the two…The R–Q view of
> variables could be called the pronoun view of variables. Professor Quine has compared the
> use of variables with many uses of pronouns in English; however, he views pronouns as more
> fundamental than nouns…On the R–Q view, quantifiers are used to talk about all entities
> or some entity…On the F–L view, a variable is seen as a replacement for an expressions.
> Variables do not have ranges of values, where each value is an entity of some kind.

I am not convinced that Kearns is exactly right crediting the substitutional reading to Leśniewski;
it seems that his interpretation relies on a mistranslation. Leśniewski in the passage Kearns refers
to explains how he used quantifiers when he **"did not yet know how to operate with quantifiers."**
It is therefore not sure whether this is how he understood his formal quantifiers once he had them.
Also, he wrote "przy pewnym znaczeniu wyrazu", which translates as "for some meaning of the
expression" or "for some meaning of the word" (Leśniewski 1927, 203). Kearns translates the phrase
as "for some significant words" (which would be "dla pewnego wyrazu posiadajacego znaczenie"
or "dla pewnego znaczacego wyrazu" in Polish) and takes it to be Leśniewski's explanation of his
quantifiers. But let's put these qualms aside.

[12] I have not encountered an argument formulated exactly this way, but it seems that the tendency
can be traced back at least to Quine. In 1947 he remarks that any usage of general terms in the
context of quantification commits one to abstract objects (pp. 74–75), and he explicitly says about
quantifiers binding predicate letters:

> If we bind the schematic predicate–letters of quantification theory, we achieve a reification
> of universals which no device analogous to Fitch's is adequate to explaining away. These
> universals are entities whereof predicates may thenceforward be regarded as names; they
> may be construed as attributes or as classes …The predicate letters, when thus admitted to
> quantifiers, acquire the status of variables taking classes as values (pp. 77–78).

He is even more explicit when he discusses Leśniewski's Ontology (Quine 1952, 141). There,
he criticizes Ajdukiewicz and Leśniewski for not attaching sufficient significance to the fact that
"the variables which have been said to stand in places appropriate to general terms are subjected
in Leśniewski's theory to quantification." This theory, according to Quine, "…surely commits
Leśniewski to a realm of values of his variables of quantification; and all his would-be general
terms must be viewed as naming these values singly." Quite rhetorically Quine claims that if this
quantification is not to be read as committing Leśniewski to classes, then he is "at a loss to imagine

For the sake of argument I will grant [SA1-2]. I will, however, later take an issue with [SA4], and by dealing with it, also with [SA3].

Why exactly should we accept [SA4]? When it comes to problems with the nominalistic acceptability of the substitutional interpretation of Leśniewski's quantifiers, Küng and Canty (1970) is a classic.[13] The authors go through a certain variety of approaches to the problem and there are some bells and whistles that we will soon look into. The main gist, however, is that no matter whether we think of names we can substitute as existing inscriptions, or whether we think of them as possible terms that can be introduced by means of a definition, there can be at most countably many of those. In the first case, because they would be finite sequences over a finite alphabet (this is how a nominalist should think of names, arguably). In the second case, because definitions are finite sequences over a finite alphabet. This leads to a problem when we want to emulate set-theoretic second-order quantification over an infinite domain, for there are uncountably many subsets of each such domain, and thus we will run out of names supposed to go proxy for subsets. This indicates that it will be hard to run certain mathematical theories. For instance, it will be very hard to mimic the quantification over real numbers using this sort of substitutional reading. Also, even if the domain of objects is denumerable, the substitutional interpretation will give us a different result than the set-theoretic one, because there will be more subsets of the domain than possible names.

Let us call the arguments against any substitutional interpretation of higher-order (or plural) quantification which in one way or another relies on the claim that since

---

(Footnote 12 continued)

wherein such commitment even on the part of a professing Platonist can consist". Quine's argument is, however, slightly different from the one I gave. He is already assuming that being subjected to quantification automatically carries ontological commitment, and instead of explicitly discussing the need for semantics and how what semantics we choose can impact our ontological commitment he insists that sentences should be put in their standard first-order form. Quine (1970, 66–67) contains yet another argument. In ordinary first-order quantifications (like the one expressed by '$(\exists x)(x$ walks$)$') the open sentence displays a variable in a position where a name could stand. The quantified sentences are not read as quantifying over names: they rather quantify over objects that could be named by a name placed in the position of a variable. Similarly when we bind predicate variables, we "treat predicate positions suddenly as name positions, and hence …treat predicates as names of entities of some sort." This argument has been dealt with by Boolos (1998b, 37–39), who points out that to put a predicate variable in a quantifier may require that the variable be treated as having a range but we do not have to treat predicate positions as name positions just because we do so for ordinary individual variables.

A train of thought which more explicitly relies on the relevance of the choice of a semantics can be found in Burgess (2004, 217):

> …if a model theory has not yet been developed of a given logical notion, it may be alleged that the notion is 'meaningless' because it lacks a 'semantics'. On the other hand, once a model theory *has* been developed for a given logical notion it may be alleged that problematic 'ontological commitments' are implicit in the use of the notion…

[13] The authors argue that Ontology cannot be given substitutional semantics, and since they focus on the part of Leśniewski's Ontology which resembles QNL (i.e. they discuss quantification over name variables only), their arguments, if compelling, *mutatis mutandi* apply here.

the class of substituents is at most countable (and this constitutes a problem because we do not have any upper limit on the size of the domain of objects to be named) **size limitation objections**.

[SA4] is motivated by the size limitation objection and relies on the claim that every nominalistically acceptable approach to substitutional semantics for QNL puts the countability requirement on the set of names that can be substituted. In what follows, however, I will argue that there is a way of circumventing the intuitions underlying [SA4]. I will develop a modal-substitutional semantics for plural quantification to argue that the disjunction in [SA3] is not as exhaustive as it may initially seem.

8.4 Challenges to the Substitutional Reading

Küng and Canty (1970) argue that Ontology cannot be given substitutional semantics by discussing some substitutional interpretations which they find initially plausible and reject them one by one. Here is how they proceed.

Actual inscriptions

Suppose the quantification ranges over actual inscriptions. For instance:

$$\forall a, b\, (a\, \varepsilon\, b \to \exists c\, c\, \varepsilon\, a) \tag{8.3}$$

on this interpretation reads:

Reading 1 For any two name inscriptions a, b, if a sentence inscription in which a is subject and b predicate is true, then there is some name inscription c, contained in a true sentence in which c is the subject and a name equiform to a is the predicate.

Note that some sort of equiformity relation between inscriptions is necessary here. This point might be made clearer when we notice that $\forall a\, (a\, \varepsilon\, a \lor \neg a\, \varepsilon\, a)$ should come out substitutionally valid. If every occurrence of a in the formula was to stand for one and the same inscription the sentence would be true under no substitution, for no single token can occur at more than one place.[14]

It also seems that the way the authors read the formula is still unsuccessful. For say a name a does not occur in any sentence. Then, there is no such a thing as **the** sentence in which a is a subject. "Stand-alone" names would lie outside the range of quantification on this reading. Perhaps this can be fixed by the following reading:

---

[14] We have to be aware of the usual qualms about equiformity (the authors do not define the notion and take it to be primitive). If the quantified variables range over equiform names indistinctively, the language cannot contain equiform names which differ in their reference (which means that natural languages are not good candidates). To avoid this, one would have to construct a more fine-grained relation of equiformity, where only those "geometrically equiform" names which have the same "lexical entries" can be really equiform. But this would indicate that the notion of equiformity has some semantical component. Let's not dwell on this now, suppose that we have answered the qualms about equiformity. Then, still, as the authors claim, we run into troubles.

Reading 2 For any two name inscriptions a and b, if a sentence inscription $a_1\ \varepsilon\ b_1$ is true and a_1 is equiform to a and b_1 is equiform to b, then there is a name inscription c such that there is a true sentence $c\ \varepsilon\ a_2$ and a_2 is equiform to a.

This is still not exactly right. The fact that a certain sentence inscription logically implies another sentence inscription is made contingent on the existence of yet another sentence: the entailment between $a\ \varepsilon\ b$ and $\exists c\ c\ \varepsilon\ a$ is made contingent upon the existence of a certain $c\ \varepsilon\ a_2$. Rather, the intuition is that, for instance, 'Russell ε philosopher' entails '$\exists c\ c\ \varepsilon$ Russell', no matter whether there *actually* exists a true sentence which would result from '$c\ \varepsilon$ Russell' by substituting a name for c.

A modification of reading 2 would not require the existence of sentences. It would be enough that name inscriptions exist and refer. For instance, (8.3) would be rendered as:

Reading 3 For any name inscriptions a and b, if a refers to a unique object which is among the objects to which b refers, then there is an actual name inscription c such that it refers to a unique object which is one of the objects to which a refers.

On this reading, the formula would be valid (because the existence claim in the consequent is warranted by the existence of a itself). This builds semantics into the picture and fixes some of the difficulties discussed above.

Still, the problem is that any theory which requires that the domain be at least denumerable (like Peano arithmetic, for instance) will fail to be modeled on this reading since the number of existing inscriptions is very likely finite. Since a very important theoretical purpose of logical systems like Ontology (or systems based on the logic of plurals) is to develop a formalization of mathematical discourse, if we give QNL a substitutional semantics of this sort it will fall short of providing us with the needed framework.

Logically possible inscriptions

Having abandoned the actualist substitutional reading Küng and Canty (1970) then consider the idea that Leśniewskian quantifiers range over "logically possible inscriptions" (p. 174).

On this reading, an existentially quantified statement would be satisfied iff there were a logically possible inscription which would satisfy the requirements put on it by the content of the formula. As they quite correctly notice, there is a problem with explicating the notion of logically possible inscriptions. They make a valid negative point that logical possibility should not be identified with physical possibility.[15] As for the positive part, they focus first on the reading on which logically possible names

---

[15] In a way, Leśniewski's metalogic seems to require a modal approach as well. Kearns (1962, 11) says about Leśniewski's terminological explanations (which I comment on in sect. 6.4): 'The T.E.s are like blueprints—they present the rules for as many systems as one cares to construct. For Lesniewski, a formal system exists in space and time, and it contains just those results which have actually been deduced. This means that there are no unproved theorems of a formal system. However, such a claim does not make proofs of completeness and consistency either trivial or senseless. For

are those that either actually exist or can be introduced according to the Leśniewskian rule of definition.[16] The authors, however, do not find this initially plausible reading very satisfactory:

> But in ontology proper only two name constants are definable, the empty name and the universal name, just as in Boolean algebra only two non-equivalent constants are distinguishable (p. 177).

Depending on how one distinguishes name constants, the claim may be true or false. On one reading, there are only two constants definable (suppose we reasonably extend S-semantics to allow constants to denote) if in any S-model any definable constant will denote one of certain two subsets of the domain, given that no constants can be taken as primitive. For instance, if in any model any defined constant will denote the empty set or the whole domain of the model, we can (in the sense just specified) say that at most two constants are definable. In this sense, the above claim is true.

On the other hand, if what we mean by 'different constants' is 'constants which are not logically equivalent', that is, constants such that there is a model where they denote different subsets of the domain, there are more than two constants. It is the second meaning of 'difference between constants' that led Rickey (1985, 186) to criticize Küng and Canty (1970). Let us take a look at his argument.

He suggests that there are other, non-equivalent constants that can be defined in Ontology (which are also definable in QNL). His example is:

$$\forall a \, (a \, \varepsilon \, 2^+ \equiv a \, \varepsilon \, a \wedge \exists b, c \, (a \, \varepsilon \, b \wedge a \, \varepsilon \, c \wedge \neg \, b \circ c)) \tag{8.4}$$

where $b \circ c$ stands for $\forall d \, (d \, \varepsilon \, b \equiv d \, \varepsilon \, c)$. Further, he claims the following. In a model which contains exactly three objects (say, 1, 2, 3) where b refers to two objects: 1 and 2, and c refers to two objects: 2 and 3, 2^+ is a constant different from both the empty and the full name, because it denotes only one object, 2.

The initial worry about the fact that only Λ and V are definable was that on this sort of substitutional reading the satisfaction of complex formulas will depend only on what is true or false about the empty set or about the whole domain, that is,

---

these must be understood as proofs about all possible well-formed formulas or theorems, rather than (Footnote 15 continued)
as about some ideal individuals. The problem of potentiality, or possibility, is a different problem from that relating to abstract entities. Potential objects and events are quite as individual as actual ones. To say that every well-formed formula which can be constructed will either be provable or contradictory is not to say anything which commits one to recognizing **the** formal system containing **the** theorems.' [the emphasis is Kearns's]

[16] In QNL we could introduce a rule of definition in the following form. First we extend the alphabet by a denumerable number of name constants $\delta_1, \delta_2, \ldots$ and modify the syntax accordingly. Next, we enumerate all QNL formulas which do not contain name constants and contain exactly one free variable: $\phi_1, \phi_2, \ldots$. Then we introduce as axioms all formulas of the form:

$$\forall a \, [a \, \varepsilon \, \delta_i \equiv a \, \varepsilon \, a \wedge \phi_i(a)]$$

and in any model we take δ_i to denote the set of all and only those objects whose singular names would satisfy ϕ_i.

only the empty set or the whole domain would, in a sense, lie in the range of name variables (in a sense, this is the semantical version of the claim that there are not enough constants). The question is: do constants introduced according to Rickey's recipe ($2^+, 3^+, 4^+$, etc.) fix the problem?

One worry is that Rickey in his claim already assumes he understands quantification and knows that b or c are legitimate subsitutends. But the approach is only about to provide constants! Let us put this worry aside, what objects exactly fall under 2^+ in Rickey's three-element model?

In a purely substitutional interpretation it is somewhat hard to decide what this question means,[17] but a plausible shot might be this. An object x from the model described by Rickey falls under 2^+ in that model if and only if, if we introduce a singular term a denoting x, then the description of the model makes $a \, \varepsilon \, 2^+$ true.

So say we start with a language with three singular terms: '1', '2' and '3' denoting objects 1, 2 and 3 respectively, and suppose we have constants b and c with reference as specified above. Take 1. Certainly $1 \, \varepsilon \, 1$. Also, $1 \, \varepsilon \, b$ and $1 \, \varepsilon \, V$. Thus $\exists b, c \, (a \, \varepsilon \, b \, \wedge \, a \, \varepsilon \, c \, \wedge \, \neg \, b \circ c)$. Hence $1 \, \varepsilon \, 2^+$. Similar arguments work for 2 and 3. Thus, in Rickey's model 2^+ has the same denotation as V. In general, any constant of the kind described by Rickey of the form n^+ (where is a natural number) will denote the same as V in any model which has at least n elements and the same as Λ in any model which has less than n elements. Extending the language with logically non-equivalent constants defined in accordance with Rickey's recipe does not provide us with a way out.

Küng and Canty propose a yet wider notion of logical possibility of constants. First, they remark that even though certain constants cannot be defined in the language of Ontology, they can be introduced axiomatically.

> ...new constants can be added to Ontology not only by means of definitions, but also by means of new axioms (Küng and Canty 1970, 177)

This of course is a little bit tricky, because it is not obvious what conditions have to be put on an axiomatic system in order for it to **really** introduce a new constant. If what one cares about is only that introduction of new constants preserves consistency it is enough to add axioms saying that a new constant is not coextensive with any other constant. If, on the other hand, one believes that a new constant is really introduced if the axioms together with rules of inference determine the semantical interpretation of the new expression, things can get fairly complex.

Fortunately we will not have to get into this issue too far, because, arguably, no matter how exactly axioms will introduce new constants, we still will run out of constants. We may safely assume that any constant would be introduced my means of a single formula or a single axiom schema (say, the conjunction of those axioms and/or axiom schemas which would introduce it). But this implies there are at most denumerably many formulas (or schemas) that introduce a new constant.

---

[17] The problem here is that on the classical reading "substitutional quantification is not grounded in assignments of objects at all" (Marcus 1972, 248). Rickey, on the other hand, quite interestingly, wants to have his pie and eat it: to interpret quantification substitutionally but assign denotation to substituents. It is not impossible (in fact, I will be trying to do something similar), but this makes the semantics far from being purely substitutional.

This indicates that it will be hard to run certain mathematical theories. For instance, it will be very hard to mimic the quantification over real numbers using this sort of substitutional reading. Also, even if the domain of objects is denumerable, the substitutional interpretation will give us a different result than the set-theoretic one, because there will be more subsets of the domain than possible names.

Here, Küng and Canty end the list of what they consider acceptable approaches to the substitutional quantification[18] and go on to suggest that...

> one must further consider the set of all subsets of that domain. Here, one assigns to each constant from the category of names some element from this set: call it the extension of the name to which it is assigned. Under any interpretation of ontology, every name is to have some extensions ...the names do not denote their extensions, they merely have their extensions (p. 178).

and this is where we leave them. The problem is that extensions thus understood are *prima facie* nominalistically unacceptable, and the authors do not provide any argument to the effect that they are not.[19]

---

[18] I discussed their arguments because they focus on arguments against substitutional interpretation of higher-order logics specifically and they cover most of the interpretations that usually come to one's mind. As for general qualms about substitutional interpretation I believe they have been dealt with by Kripke (1976). I will briefly discuss an objection raised against the substitutional interpretation (not necessarily higher-order) by Tomberlin (1997). The argument is intriguing, and it is interesting to see how languages based on QNL avoid the difficulty. It goes (p. 160): On the substitutional reading a sentence of the form $\forall x\, F(x)$ is true iff all substitution instances of Fx are true (where F is given a fixed interpretation). But 'the unique non-F' is a legitimate substituent (terms do not have to be non-empty to be rightful substituents). Unfortunately, the sentence 'F(the unique non-F)' will always be false. Hence, the sentence '$\forall x\, F(x)$' will never be true, because it will always have at least one false substitution instance.

One might object by denying that definite descriptions are legitimate substituents (because they are incomplete symbols). Tomberlin has a rejoinder here: he can introduce 'Scotty' as a non-referential name, and fix its denotation (if it were to have one) as the one and only individual that is not F and run the argument with 'Scotty' instead of the definite description.

The problem does not arise for QNL even if we admit definite descriptions as rightful substituents. First, how do we express the sentence that all objects have a certain property, or, when we speak QNL that all objects fall under a certain name, say a? Certainly not by '$\forall b\, b\,\varepsilon\,a$', because this will always be false (the substitution instance '$\Lambda\,\varepsilon\,a$' is false). Rather, we say that any **object** is a, '$\forall b\,(b\,\varepsilon\,b \to b\,\varepsilon\,a)$'. But consider the substitution instance that we get when we substitute 'the unique non-a' for b:

If the unique non-a is the unique non-a, then the unique non-a is a.

If there are no objects which are not a (or if there are at least two such objects), the descriptions fails to denote and the sentence comes out vacuously true. If, on the other hand, there is a unique object which is not a, then the sentence also gets the correct truth-value, for it is true that the unique non-a is the unique non-a but false that it is a.

[19] They remark that "A Leśniewskian system ...has the advantage that it is at least as powerful as any Russellian system without admitting sets into the universe of discourse; that is, without putting, say, real things and sets ...into the same universe. The sets are here recognized as purely linguistic entities, which play their role as extensions in the range of quantification, not in the universe of discourse." (p. 179) but this looks more like mere verbalism than like a real explanation. The reader is still left with the idea that sets have to be in the range in quantification and has no idea as to how they are to be explained away by saying that they are 'purely linguistic entities'.

8.5 Simons's Combinatorial Semantics

Another account of Leśniewski's quantifiers has been developed by Simons (1985), who suggested that we can think of Leśniewskian quantifiers as ranging over **ways of meaning** (a phrase he coined, henceforth 'WOMs'). He did not define the notion, but he made important moves to clarify it. The intuition is that the WOM of an expression is that part of the semantical character of this expression which contributes to the truth of complex sentences (of the extensional language of Ontology). Two examples: the WOM of 'Rafal Urbaniak' is that it refers to only one object, me. The WOM of the Boolean conjunction is that only when applied to two true sentences it yields a true sentence.

First of all, we are told, in the present context we have to consider only extensional aspects of WOMs. Hence his first postulate:

(8.5) Coextensive expressions mean in the same way.

Also, the grammatical differences are supposed to "represent" differences in reference (denotation):

(8.6) No two expressions of different semantic category can mean in the same way.[20]

In other words, two expressions which **could** mean in the same way, possess the same **mode of meaning** and expressions in different sematic category never have the same mode of meaning. Two expressions can have the same mode of meaning and yet have different WOMs.

(8.7) Expressions have the same mode of meaning iff they are of the same semantic category.

(8.7) is not a definition. It only specifies that the way Ontology is constructed its grammar mirrors the distinction between different modes of meaning. Then we have two principles associated with compositionality:

(8.8) The WOM of a complex expression is determined by the WOMs of its constituents.

(8.9) The mode of meaning of a functor expression is determined by the modes of meaning of its arguments and values.

Now Simons has to specify the mode of meaning of expressions of primitive syntactic categories:

(8.10) The mode of meaning of sentences is to be true or false.

(8.11) The mode of meaning of names is to designate.

Among the WOMs that names can have, Simons distinguishes certain **submodes**; there are lots of submodes but those interesting from the present perspective are designating no/at least n/exactly n/at most n/all objects, etc.

---

[20] Simons is using Leśniewski's terminology, so he speaks of 'semantic category'. In Ontology, however, the category of an expression is determined syntactically.

An example. Suppose we consider an expression which syntactically is a two-place propositional connective (or suppose we are considering a variable of the same syntactic category and ask ourselves what the range of quantification is). How many WOMs can it have? Well, a WOM of a propositional connective is specified by its output for each possible input, we are told. There are 2^n possible inputs, and since there are two possible outputs, there are 2^{2^n} possible WOMs of such a propositional connective.

The WOM of a name constant is specified by determining what objects it refers to. Simons represents WOMs of expressions of propositional logic by means of tables (and this is doable, because they would always be finite objects), and suggests that this method can be somehow extended to WOMs that names have. For this purpose he introduces **designation tables**. For instance, if there are two objects in the domain: A, B there are four ways a name could mean and we can represent them by $a, b, -,$ and ab respectively:

Does it denote	A?	B?
a	Yes	No
b	No	Yes
$-$	No	No
ab	Yes	Yes

Suppose we now want to specify the way ε means in that model based on the WOMs of its two arguments:

Inputs		Output
WOM of first argument	WOM of second argument	
$-$	$-$	F
$-$	a	F
$-$	b	F
$-$	ab	F
a	$-$	F
a	a	T
a	b	F
a	ab	T
b	$-$	F
b	a	F
b	b	T
b	ab	T
ab	$-$	F
ab	a	F
ab	b	F
ab	ab	F

The idea is that quantification ranges over WOMs (e.g. if a is the only free variable in $\phi(a)$, the sentence $\forall a \, \phi(a)$ is satisfied in a model if and only if for any WOM a can have $\phi(a)$ is true in that model), Simons suggests, and such quantification does not commit one to the existence of objects which are WOMs. Simons remarks: "Note again that the inputs and outputs of these semantic tables are not expressions or any other objects connected directly with a particular language. They are ways of meaning" (Simons 1985, 208). Simons also argues that WOMs cannot be actual objects, because "there are more ways of meaning than the objects to be named" (p. 212).

This argument from cardinality is a bit hasty, though. It seems that a certain ambiguity is involved. In one sense an object has to be an element of the domain of things that names can name. On this reading, we probably do not want to treat WOMs as objects, that is, we do not want to consider possible names which name WOMs as admissible substituents for basic name variables to avoid diagonal arguments.

In the other sense (which probably is more essential here) WOMs are actual objects if they really exist, no matter whether we can put them in one single domain of quantification or not. The claim that WOMs are not objects in the first sense (likely true) does not entail that they are not objects in the second sense. The first claim is only about a certain limit on what names can refer to in a certain (possibly formal) language. The second claim is general and metaphysical, and it is this second sense which is relevant for questions of ontological commitment. (In a sense, what is needed for the passage from the former to the latter is the unobvious assumption that all objects can be thrown together in one big domain of truly universal quantification.)

Say we agree that WOMs are not objects. It is still not clear what they are. The postulates given by Simons or the fact that some of them (those which are generated from a finite domain of "urelements") can be represented by tables do not provide a philosophical explanation of their metaphysical status.[21]

Even if we assume that tables adequately represent ways names could mean with respect to a model, the method does not make it possible to express the meaning of ε. The best that a table can do is to express the way ε, means with respect to a model, and this works also only in finite models. What we want to have is a general description of what is common to all the WOMs of ε in all models, and this cannot be represented by a table. Thus, Simons's approach, as he admits, offers only an elucidation, not a full-blown account of the notions involved. While sometimes this is the best we can do, it might be worthwhile to see if there are more elaborate philosophical accounts on the market.

8.6 Rickey's Natural Models

After a simple reformulation (accommodating it to QNL), Rickey's natural semantics (Rickey 1985) is a model theory for QNL which, he believed, does not commit one to sets when one uses the system. The basic idea is that we take names to belong

---

[21] There is also a slight technical difficulty here. Unless we represent tables set-theoretically, a formal semantics which would employ tables that list objects, inputs and outputs would be quite unmanageable.

to a distinct set of objects existing in the model and denotation relation to be given together with the world.

More specifically, Rickey's **natural model** is a tuple $\langle i, n, \sim \rangle$, where i is a (possibly empty) set of individuals in the world, n is a non-empty set of names in the world and $\sim \subseteq n \times i$. A **natural interpretation** of a formula ϕ is a natural model together with a valuation function v which maps name variables occurring in ϕ into n.[22] Let $I = \langle i, n, \sim \rangle$. The definition of satisfaction (relative to a valuation function) is:

$$\langle I, v \rangle \models a \, \varepsilon \, b \ \text{iff} \ \exists!_{x \in i} v(a) \sim x \wedge \exists_{x \in i}(v(a) \sim x \wedge v(b) \sim x) \qquad (8.12)$$

$$\langle I, v \rangle \models \neg \phi \ \text{iff} \ \langle I, v \rangle \not\models \phi$$

$$\langle I, v \rangle \models \phi \wedge \psi \ \text{iff} \ \langle I, v \rangle \models \phi \text{ and } \langle I, v \rangle \models \psi$$

$$\langle I, v \rangle \models \forall a \, \phi(a) \ \text{iff} \ \langle I, v' \rangle \models \phi(b)$$

where b is a variable which does not occur in ϕ, and

v' extends v by mapping b into n as well.

Rickey constructed interpretations of this sort to criticize Küng and Canty (1970) by showing that substitutional interpretations can have an infinite domain of individuals and nevertheless the domain of names does not have to be uncountable to provide sufficient "substitutional" foundation for higher-order quantification. For indeed, even if the domain of $\langle I, v \rangle$ is infinite, there are uncountably many $\langle I, v' \rangle$ to do the right job in cooperation with clause (8.12).

A technical remark first. As the semantics stands, it does not avoid the difficulties raised by Küng and Canty. For even an extended valuation function v' of (8.12) will map new variables into the same old n with the $\sim$ already fixed. (Thus, quantification will be in fact restricted to extensions of those names which already are in the original model). This is rather clearly a feature that Küng and Canty would find problematic. In other words, what we get is a semantics equivalent to non-standard set-theoretic semantics for higher-order logic, and that means that we will not have certain properties that higher-order logic usually is expected to have (incompleteness, categoricity etc.).

Also, the satisfaction conditions do not seem right if $\sim$ is not allowed to vary. If, for example, in I all names are $\sim$-related to one and the same object, no extension of v to a new name variable will allow a name to refer to something else than that object.[23]

Perhaps it would agree more with Rickey's intentions to say that $\forall a \, \phi(a)$ is satisfied in a model $I = \langle i, n, \sim \rangle$ under a valuation v if and only if $\phi(a)$ is satisfied in

---

[22] This is the reason why Rickey postulated that n be non-empty. He needed at least one object to map name variables to.

[23] I owe this remark to Richard Zach.

any model $I' = \langle i, n', \sim' \rangle$ (where $n' = n \cup \{y\}$ and y is a new element of n' and $\sim'$ extends $\sim$) under the valuation v' that assigns this new element y of n' to a.[24]

Rickey did not give any argument to the effect that QNL with his semantics does not carry ontological commitment to sets. A plausible case for this claim, however, can be made. First, the quantification itself does not bring a commitment to sets. It is substitutional and names are individual objects taken to exist in a world pretty much just like individuals. Second, the way that names contribute to the truth conditions of atomic sentences is non-committal either. **Names** on this view **do not refer to objects via sets**. They simply **refer to objects directly**. Saying that a name refers to objects (e.g. that 'dog' refers to dogs) is not the same as saying that it names a unique set whose members fall under this name (e.g. that 'dog' names a unique set whose all and only elements are dogs and dogs fall under the name 'dog' in virtue of belonging to this set).[25]

8.7 Relational Semantics for Plural Quantification

Borrowing intuitions from Simons and Rickey, let us try to develop a nominalistically acceptable interpretation of plural quantifiers.

In the relational semantics for QNL quantifiers range over certain **ways names could be**. Relational models defined below will introduce the notion formally in a fairly well-known framework. This will allows us to get a better handle on the semantics and hopefully provide us with a better understanding of this notion.

Suppose we start with a world which for our purposes is devoid of linguistic objects (especially names). The content of this world is our domain of (already existing) extralinguistic individuals, call it the **bare world**. In such a world, it is possible to introduce names which would either be empty or refer to one or more already existing objects. In this setting we can take a possible world to consist of two sorts of objects: bare individuals (those which do not name anything) and names that either do not name anything or name (one or more) extralinguistic individuals. The situation in a possible world determines what reference the names that exist in it have (if there are any). The bare world can be also interpreted as such a possible world, only (at least) one of the sorts would be empty. Starting with the bare world we can subsequently extend its repertoire of names by introducing (countably many) new names that refer to bare individuals. Let us idealize here: we are not putting any restrictions on which individuals can be named by a name, we assume that any bare individual (or any bare individuals) is (are) nameable. This gives raise to the so-called **naming structure**:

---

[24] Another option would be to avoid extending the set of names and just vary $\sim$. This would yield the same result. I prefer to extend the set of names in order to make this semantics more similar to the one I will talk about soon.

[25] A similar point has been made by Boolos (1998a), where the reader can look for a more emphatic and detailed explanation of the difference.

Definition 8.1. A **naming structure** is a tuple $\langle I, W \rangle$ where I is a set (of bare individuals) and W is a set of possible worlds. A **possible world** is a tuple $\langle N, \delta \rangle$ where $I \cap N = \emptyset$ and $\delta \subseteq N \times I$. A **bare world** is the possible world with $N = \emptyset$. The following conditions all have to be satisfied:

(i) $B = \langle \emptyset, \emptyset \rangle \in W$ (i.e. the naming structure contains a bare world).
(ii) For any $B \neq w = \langle N, \delta \rangle$, N is countable and $N \neq \emptyset$.

The accessibility relation on possible worlds is defined by the following condition. Let $w = \langle N, \delta \rangle$, $w' = \langle N', \delta' \rangle$. Rww' if and only if $N \subset N'$ and $\{\langle x, y \rangle \mid x \in N \wedge \langle x, y \rangle \in \delta'\} = \delta$ (i.e. $\delta' \cap N \times I = \delta$). ⊣

If **M** is a naming structure and I is the set of its bare individuals, I say that I **underlies M** or that **M** is **based on** I. If $\mathbf{M} = \langle I, W \rangle$ and $w \in W$ I will sometimes write $w \in \mathbf{M}$.

A few words of explanation here. First, we start "constructing" a naming structure with a bare world. This bare world together with I represents the situation where we have a set of objects that can be named but we have not introduced any names yet. Since in principle (unless one has very specific religious beliefs) there is no reason to believe in unnameable individuals, the I in the bare worlds is simply the domain of objects.

We are considering ways individuals which are not names could be named. This means that introducing new names does not change the domain of individuals (that is why I is a set in the naming structure and does not vary with possible worlds). Also, this indicates that the only way we get from one possible world to another accessible world is by **extending** the repertoire of available names (so N has to be a proper subset of N').

On the nominalist reading, quite plausibly, names are finite sequences of symbols (or phonemes) from a finite alphabet. This means that in any possible world there can be at most countably many names (hence the requirement that N be countable). Moreover, the basic idea is that by "going" to another possible world we are **extending** what we already have; **not changing** the ways that names that we already have are. That is, we can add a new name and take it to refer to such-and-such objects, but the reference of the already existing names cannot be changed. In a sense, names are thought of as given with reference. Hence the denotation relation in an accessible world has to agree with the initial denotation relation on all names that already existed before we extended the set of names.

This sort of structure will not ensure yet that every intuitively possible "way of meaning" will have a representation in a model. To do this, we have to require that for any subset of the domain of individuals in a possible world there be an accessible possible world in which there is a name which denotes all and only its elements.

Definition 8.2. Let $\langle N, \delta \rangle = w \in W$. A naming structure $\mathbf{M} = \langle I, W \rangle$ is w **-complete** if and only if:

$$\forall A \subseteq I \, \exists w' = \langle N', \delta' \rangle \, (Rww' \wedge \exists x \in N' \, \forall y \in I \, (\delta'(x, y) \equiv y \in A))$$

$\mathbf{M} = \langle I, W \rangle$ is **complete** iff for any $w \in W$ $\mathbf{M}$ is w-complete. $\dashv$

The basic idea is that a naming structure is w-complete iff for any set of individuals existing in this world, there is a world accessible from w where a name which names elements of this set exists. In other words, it is w-complete if it models the full range of ways names of objects from w could be.

The reason why B-completeness is not sufficient is that the class of B-complete naming structures would agree with classical set-theoretic semantics only on sentences with only one quantifier. In a naming structure, the iteration of quantifiers carries us deeper and deeper into the structure.

I can now explain how quantification in QNL is supposed to range over ways names could be (or refer). I will be evaluating sentences in a name structure by evaluating them in its bare world. First, we have to define satisfaction of a formula at a world in a naming structure.

Definition 8.3. An **M-interpretation** is a triple $\langle \mathbf{M}, w, v \rangle$, where $\mathbf{M}$ is a naming structure, $w = \langle N, \delta \rangle$ is a possible world in $\mathbf{M}$ and v either assigns to every variable in QNL an element of N, if $N \neq \emptyset$, or is the empty function on the set of variables of QNL otherwise. If $\mathbf{M}$ is a complete naming structure, then we say that this M-interpretation is complete.

Satisfaction of formulas under **M-interpretations** is defined as follows.

Definition 8.4. Let $\langle \mathbf{M}, w, v \rangle$ be an M-interpretation, $w = \langle N, \delta \rangle$. Also, let a and b be QNL-variables and ϕ and ψ be QNL-formulas. The relevant satisfaction clauses are:

$\langle \mathbf{M}, w, v \rangle \models_\diamond a \, \varepsilon \, b$ iff $v(a)$ and $v(b)$ are defined, $\exists!_{x \in I} \langle v(a), x \rangle \in \delta$,

and $\exists y \in I (\langle v(a), y \rangle \in \delta \wedge \langle v(b), y \rangle \in \delta)$.

$\langle \mathbf{M}, w, v \rangle \models_\diamond \neg \phi$ iff v is not the empty function and $\langle \mathbf{M}, w, v \rangle \not\models \phi$.

$\langle \mathbf{M}, w, v \rangle \models_\diamond \exists a \, \phi$ iff for some $w' \in M$, Rww' and $\langle \mathbf{M}, w', v' \rangle \models \phi$

where v' differs from v at most in what it assigns to a.

(I skip the clause for conjunction because it is fairly standard, but I give the clause for negation because it is slightly different from the usual formulation.)

The basic idea is that we treat the class of names that exist in a certain world as the class of substituends when we evaluate a formula in this world. Since the bare world does not contain names the notion of satisfaction in the bare world as applied to open formulas is not very fascinating. No formula containing free variables will be satisfied at the bare world. The situation changes when it comes to evaluating **sentences** in bare worlds. First, let us say that **a sentence is true in a naming structure M** if and only if it is satisfied in its bare world under any valuation. **A sentence is M-valid** if and only if it is true in any naming structure. A sentence is **M-complete valid** if it is true in any complete naming structure.

Theorem 8.1. *The following correspondence holds:*

(a) *For any set-theoretic QNL model there is a complete M-model which agrees with it on all formulas.*

(b) *For any complete M-model there is a set-theoretic QNL model which agrees with it on all formulas.*

Argument By model construction:

Ad (a). Suppose a formula ϕ fails at an S-model $M = \langle D, In \rangle$, let $a_1, \ldots a_k$ be all variables that occur in ϕ. Take a naming structure w with $I = D$. Since the structure is complete, there exists a possible world $w = \langle N, \delta \rangle$ such that for any $A \subseteq D$, if for some i, $In(a_i) = A$ then there is a $y \in N$ such that for all $x \in I$, $\delta(y, x) \equiv x \in A$. If $In(a_i) = A$ let $v(a_i)$ be an arbitrary $y \in N$ for which this condition ($\forall x \in I$ ($\delta(y, x) \equiv x \in A$)) holds. Then also ϕ fails at this possible world under v. The proof (that M and w agree on ϕ) goes by induction. If, on the other hand, ϕ does not fail at M, then it also does not fail at any naming structure w with $I = D$. If it did, then using the procedure described in (b) below we could show that it would also fail at the original S-model.

Ad (b). I will just show, given a non-bare world in a naming structure and a valuation function, how to construct a corresponding S-model that satisfies exactly those formulas that are satisfied at that world under that valuation. Take a $w = \langle N, \delta \rangle$ in a naming structure based on I. Let v map name variables $a_1, a_2, \ldots$ into N. To obtain an equivalent S-model $w' = \langle D, In \rangle$ take $D = I$ and let $In(a_i) = \{x \in I \mid \delta(v(a_i), x)\}$ for any a_i. The proof that two such models satisfy the same formulas is by induction.   □

The following observations are in order. First, *modulo* this relational semantics, QNL has an intuitive translation into a two-sorted first-order quantified modal logic with one additional primitive binary operator representing the naming relation, which supports the idea that the commitment of QNL doesn't have to go beyond that of first-order modal logic (the translation is rather straightforward). Second, the size limitation objection does not apply here. Quantifiers are interpreted as ranging over possible names, but not over possible names from one particular possible world but rather over names that belong to the union of all sets of names from all accessible possible worlds. The initial plausibility of the objection results from the ambiguity between the following two readings:

[R1] It is possible that for every subset of the domain there is a name which names all and only those objects which are elements of this subset.

[R2] For any subset of the domain it is possible that there is a name which names all and only all its elements.

[R2] might hold even if [R1] fails. For instance, suppose the domain consists of all real numbers. It is false that there is a possible world in which all elements of the domain are named by individual terms, because there are not enough names in any particular possible world. However, this does not mean that there are unnameable real numbers. Quite the contrary, every real number can be named in some accessible possible world. The abundance of numbers arises from the abundance of accessible possible worlds—the ways names could be.

8.8 Philosophical Qualms

Let's take a look at certain philosophical qualms about the nominalistic acceptability of this approach. The first objection focuses on the ontological commitment of the possible-world discourse. The second one emphasizes that the modal reconstructions fail to avoid the epistemological difficulties that realism runs into. The third one points out that while giving a nominalist reconstruction, I have employed set theory in metatheory. The fourth one questions whether the approach developed in this paper can handle the size limitation objection. The firth one doubts whether the current semantics improves in any sense on the semantics of plural quantification provided by Boolos. The sixth one takes issue with the assumption of completeness. The seventh objection focuses on the so-called set-theoretic content of QNL. The eighth one turns around the claim that the multiple reference relation is more committing than singular relation. Finally, the ninth objection relies on the claim that employing the notion of 'being true of' carries nominalistically undesirable ontological commitment.

8.8.1 Modality and the Innocence of Possible Worlds

An objection can be raised that while explaining away the quantification over sets I do it at the expense of taking on the commitment to possible worlds. This sort of objection has been raised by Woleński (1992) against Chihara (1990).[26]

Responding merely that possible worlds discourse is a convenient myth[27] is not philosophically satisfactory: one needs to explain how such a myth can be useful in

---

[26] There is some similarity between the modal account I propose and Chihara's use of constructibility quantifiers. There are some differences, however. When we extend the language to higher-order names (the strategy that lies beyond the scope of the present work, cf. (Urbaniak 2008c), the result will not resemble type theory, but rather a cumulative hierarchy. Instead of introducing constructibility quantifiers explicitly I model ways names could be using standard modalities and reshaping the accessibility relation, which allows for more flexibility and captures the philosophical ideas in a more familiar framework. I also think that the framework I presented is easier to generalize to obtain a better nominalistic account of the language of set theory or arithmetic (especially the fact that mathematical terms behave like singular terms), but let's not concern ourselves with these issues here (see however Urbaniak (2010) for details).

[27] "Remember, that for me, this whole possible world structure is an elaborate myth, useful for clarifying and explaining modal notions, but a myth just the same. It would be a mistake to take this myth too seriously and imagine that we are exploring real worlds, finding there real open-sentence tokens that have puzzling features" (Chihara 1990, 60).

a philosophical account of the truth of serious theoretical statements.[28] Ultimately, the nominalist, for the success of the current enterprize, should either provide a philosophically compelling nominalistic account of the possible-world talk, or to explain the modal notions involved without reference to possible worlds. While both of these issues lie beyond the scope of this chapter, the following remarks are due.

Given the multiplicity of anti-realist approaches to possible world discourse, it is far from obvious that the nominalist is committed to abstract objects just because he employs the possible-world talk. This is especially so, because the involved notion of possibility isn't as heavily theory-laden as various other modalities. The notion of nameability does not require the existence of persons who can actually name objects, nor does it require the real existence of other possible worlds where possible names exist. Certain objects are considered nameable if by introducing a name that multiply denotes exactly those objects we would not run into a paradox, and that is all there is to it. A theory of nameability thus understood is not a *chapter in mathematical theology* (if I may borrow an expression from Boolos et al. (2002, 19)) concerned with ideal naming agents, other real worlds and so on, but rather a theory which tries to describe certain conditions under which a name that denotes certain objects can be introduced without worrying about consistency and paradoxes. Vivid metaphors which employ possible worlds and an ideal naming agent may be helpful in clarifying logical properties of the modalities involved, but otherworldly objects or the possible behavior of ideal agents are not what the theory is about, pretty much like the computability theory is not really about possible worlds where certain computations are performed. Intuitively, the interpretation of quantifiers in QNL sentences boils down to phrases like 'it is possible to introduce a name (it wouldn't lead to a paradox if we introduced a name) which refers in such-and-such a way such that…'. But this, *prima facie*, does not commit one to abstract objects.[29]

Finally, even if you're not convinced that possible-world talk is nominalistically kosher, the conditional upshot remains that if the nominalist can use such modalities, they can also make sense of higher-order quantification.

8.8.2 *Epistemology and Modality*

Shapiro (1993) suggests that the epistemological challenges to Platonism (Benacerraf 1973) apply also to modal approaches:

> At least prima facie, the epistemology of the various modal notions is more tractable than an epistemology of abstract objects like sets. The contention of this paper, however, is that this promise is not delivered. The epistemological problems facing the anti-realist programmes are just as serious and troublesome as those facing realism. Moreover, the problems are, in

---

[28] Chihara has much more to say about possible worlds and modalities (see Chihara 1998), and he makes quite a strong case for the nominalistic acceptability of possible worlds talk.

[29] A big watershed here is obviously whether the modal talk can be done entirely in terms of modal operators. This issue, however, lies beyond the scope of this already long chapter.

a sense, equivalent to those of realism. No gain is posted, and in some cases there is a loss (Shapiro 1993, 456).

The core of Shapiro's criticism seems to be this:

> I show that there are straightforward, often trivial, translations from the set-theoretic language of the realist to the proposed modal language, and vice-versa. The translations preserve warranted belief, at least, and probably truth (provided, of course, that both viewpoints are accepted, at least temporarily). Under certain conditions, the regimented languages are definitionally equivalent, in the sense that if one translates a sentence ϕ of one language into the other, and then translates the result back into the original language, the end result is equivalent (in the original system) to ϕ. The contention is that, because of these translations, neither system can claim a major epistemological advantage over the other. Any insight that the modalist claims for his system can be immediately appropriated by the realist, and vice-versa. The problem, however, lies with the "negative" consequences of the translations. The epistemological problems with realism get "translated" as well (p. 457).

Putting technical qualms about translations aside, the worry seems to be that the modal reconstruction, since it preserves theoremhood, is left with the truth of pretty much the same (*modulo* translation) set of truths to explain.

There is an important distinction to be made, if this argument is to be dealt with adequately (for simplicity let's focus on a particular theory, ZFC): the distinction between a mathematical theory whose truth is to be explained itself (ZFC itself), and the same mathematical theory together with the realist philosophical story about it (in our case, ZFC plus a Platonist account of sets, call it ZFC+).

The mathematical theory itself doesn't contain any philosophical claims about the ontological status of objects that it is about. Further philosophical story is only to be provided, if the theory is to be given a deeper philosophical understanding. Indeed, even most successful mathematicians using set-theory rarely have developed any deeper philosophical ontological considerations meant to interpret what they're doing. What's more, often different mathematicians when asked philosophical questions about set theory, either explicitly don't care or give completely different answers. This suggests that to use ZFC as a successful mathematical tool one doesn't need any deep philosophical understanding of the language of set theory, pretty much like to spill ink one doesn't have to have read and understood Austin's "Three ways of spilling ink" (Austin 1966).

Now suppose one gives a modal-nominalist account of ZFC: a translation of the language of ZFC into some (not necessarily really different) language, a certain theory (call it NZFC) such that ZFC and NZFC are the same under such translation[30] and a nominalistic philosophical account of the truth of NZFC (let's dub NZFC with this explanation NZFC+).

Is the fact that NZFC is *modulo translation* pretty much the same as ZFC a serious objection against the nominalistic acceptability of this philosophical explanation? I would like to submit, it isn't. If the mere agreement on theorems of ZFC and NZFC were to bring in the same epistemological worries as those that ZFC+ encounters, any nominalistic attempt to reconstruct ZFC would be pointless to start with. This,

---

[30] Or at least such that NZFC is a supertheory of ZFC, modulo translation.

however, seems hasty and far from obvious. Even if nominalistic reconstructions are doomed, it is unlikely that it is mere theoremhood preservation *per se* that brings doom.

If NZFC+ preserved the truth of ZFC+, that would be a serious problem for the nominalist. The thing is, it isn't supposed to do that. Quite to the contrary, the nominalist doesn't want to preserve the truth of ZFC+ minus ZFC. He explicitly disagrees with it.

In general, the philosophical significance of interpretability of formal theories is a tricky issue and it is far from obvious that a mathematical interpretability result automatically shows that any philosophical account of the interpretans has to inherit the issues encountered by the Platonist's favorite philosophical account of the *interpretandum*. For instance, it is rather hasty to claim that there is a strong philosophical connection between real analysis and geometry just because the former can be interpreted in the latter.

Perhaps, a more general worry still stands: how come that the modal notions are to be more tractable than the Platonist account, and yet yield support to the same mathematical theory? Well, this depends on what we mean by tractability.

If tractability is measured in terms of strength of mathematical theories whose truth is being philosophically explained, then sure, any good reconstruction should have (more or less) the same level of tractability. But, for the reasons described above, it is far from obvious that tractability thus understood is a measure of epistemic viability of the associated philosophical account.

If tractability is rather taken to capture our intuitions about relative viability of the associated philosophical accounts, then mere theoremhood preservation of mathematical theories doesn't have to preserve tractability, and since no mathematically tractable notion of tractability thus understood has been developed, we have to rely on our philosophical intuitions and judge these things on case-by-case basis.

That being the case, I submit that the account of modalities given in the previous subsection is philosophically less demanding than Platonism. We might have good reasons to believe that introduction of a certain name satisfying such-and-such conditions wouldn't lead to contradiction, because we've played around with various paradoxical arguments enough to figure out that as long as our naming hierarchy is grounded, we're safe. On the other hand, no amount of time spent on doing logic or mathematics seems to move us towards having better epistemic access to causally inert abstract objects as pictured by the Platonist.

8.8.3 Can We Use Set Theory in Metatheory?

A related worry shines through some of Shapiro's remarks:

> To be sure, the modal notions invoked by our antirealists do have uses in everyday (non-mathematical) language, and competent speakers of the language do have some pre-theoretic grasp of how they work. But this grasp, by itself, does not support the extensively detailed

articulation of the modal notions as they are employed by our anti-realists in their explications of mathematics. I think we do have a decent grasp of the extensive notions, but
this understanding is not pre-theoretic. Rather, it is mediated by mathematics, set theory in
particular. For example, one item concerns the relationship between model theory and the
intuitive notion of logical consequence, in all its modal, semantic, and epistemic guises. Of
course, everyone who reasons makes use of logical consequence, but some ontologists give
the notion of consequence a central role in their philosophical stories. And most contemporary philosophers have come to accept the model-theoretic explication of consequence,
which is executed in set theory, with all its ontology. The question at hand concerns the
extent to which our anti-realists are entitled to the hard won model-theoretic results (Shapiro
1993, 457).

The problem seems to be that despite the fact that we do have some pre-theoretic
understanding of the modalities at play, the nominalist still needs to use model theory
to elucidate her understanding of the consequence operation involved, and this set
theory is formulated in set-theory, which bears commitment to sets.

One way to read the argument is to take it to proceed from [C1] to [C2]:

[C1] Set theory is used (indispensably) in the metatheory of a logic.
[C2] This logic is committed to the existence of sets.

But this is not a compelling strategy. After all, set theory is also used in the
metatheory of first-order logic and no one claims that by the same token first-order
quantification commits one to the existence of sets.

On a more charitable and weaker reading, the worry is that the modal semantics for
QNL is motivated by nominalism, and it is suspicious that someone who cares about
nominalistic acceptability of QNL doesn't mind using set theory in meta-theory.
Suspicious as it may be, it is not incoherent. Someone might be a mathematical
realist in general, and claim merely that it is not with QNL that the commitment to
abstract objects kicks in. But sure, this is not a standard position either.

A more uniform response on the part of the nominalist might be that she indeed
doesn't believe in abstract objects in general. Still, there is a sense in which set-theory
is still available to her in a kind of reductio argument. If she intends to convince
someone who doesn't mind using set theory about the ontological innocence of
QNL, she is free to use set-theory, no matter what she thinks about its nominalistic
acceptability.

This strategy, although not completely flawed, is rather risky. If the nominalist
really thinks set theory carries commitment to abstract objects, then even if the
Platonist might become convinced about the innocence of QNL, the argument won't
work for the nominalist herself and won't provide her with what she herself could
take as a philosophically relevant account of QNL quantification.

Another reply on the part of the nominalist might be to say that the set theory
employed in the description is very weak, and that in fact the intuitive explanation
of the semantics might be equally well given in a metalanguage employing modally
understood QNL-like quantification instead of set-theory talk. Arguably, using intuitively understood QNL quantification in the metalanguage while giving a semantics
for QNL quantification in the object language is not worse than giving semantics for

first-order quantifiers in a metalanguage in which first-order quantification already can be expressed.[31]

Yet another strategy for the nominalist would be to rely on the distinction between a mathematical theory and the mathematical-theory-cum-philosophical-account bundle, already employed in sect. 8.8.2. On this approach, the fact that the nominalist uses ZFC (or NZFC for that matter) does not make her committed to abstract objects, as long as using NZFC and additional philosophical consideration, she is able to develop a compelling NZFC+.

Of course, this is nowhere close to getting off Neurath's ship. The point is, however, that it is philosophically respectable to use NZFC (but not ZFC+) and some philosophical ideas to explain how one can interpret NZFC (or simply ZFC) in a nominalistically acceptable way (just like it is respectable to use English and certain philosophical ideas to develop a philosophy of language that applies to English itself). There is no circularity involved here: it is ZFC *with extra philosophical content* that is supposed to deliver the nominalistic acceptability of ZFC.

- A problem with this strategy is that it ultimately hinges on the availability of a plausible nominalistic interpretation of ZFC. The *onus probandi* is on the nominalist: a nominalistic interpretation of ZFC is still to be given. I believe such an account can be given, but this issue lies beyond the content of this paper (see Urbaniak (2010) for a sketch of such a nominalistic strategy; or consider an account given by Hellman (1989)).
- Yet, it is unfair to say that the nominalist fails in providing a plausible nominalistic story about QNL quantification just because she uses ZFC in model theory. The nominalist is in the process of developing a philosophical account of ZFC using (among other things) ZFC itself. Explaining why plural quantification is non-committing is only a step towards this goal, and this explanation should be evaluated as such a step.
- One could ask: if the nominalist can use ZFC in her account of QNL quantification, why not use standard set-theoretic semantics to start with? The reason why this wouldn't work is that this wouldn't contribute to showing that QNL quantification is nominalistically acceptable. In contrast, on the current approach it is emphasized

---

[31] Here is an example of how such a description could start:

- Read 'a possible world is a tuple composed of a set of individuals, a set of name tokens, and a reference relation between these two' as 'a possible world contains individuals and name tokens which refer to individuals'.
- Read the important part of Definition 8.1 as 'To evaluate formulas, we need to be told what individuals exist and what possible worlds can be considered. Among the latter, we need to include a possible world in which no names exist, and any other possible world has to contain countably many names. One possible world can access another if the latter contains all the names that the former does (preserving what they refer to), and at least one new name'.
- To ensure completeness of our evaluation framework, we also need to assume that for any possible world w, for any individuals, there is a possible world w' in which a name exists which refers to exactly these individuals.

Developing this alternative semantics in full, alas, lies beyond the scope of this chapter.

that the reading of QNL formulas does not involve reference to sets: 'it is possible to introduce a name token' is more clearly nominalistically acceptable than 'there exists a set'.

Shapiro's emphasis on the fact that set-theoretic model theory is needed to clarify the modal notions involved points to yet another strategy which the nominalist can try: to deny this assumption. Perhaps, instead of using set-theory to clarify the modalities involved, one can rather take modalities as primitive and axiomatize them directly. While this task has not been completed in this paper, Shapiro does not provide an argument against the possibility of such an axiomatization either. Of course, this controversy will have to wait for another day, when such axiomatizations are actually proposed and their nominalistic acceptability is discussed on a case-by-case basis.

8.8.4 What About the Size Limitation Objection?

We can recall that one of the objections against the substitutional interpretation of QNL was that if the domain is infinite we will always run out of substituents for under no circumstances the existence of uncountable infinity of names is possible. Naming structures by definition can only contain possible worlds with a countable number of names. It does not follow, however, that we will run out of possible names if the bare world is infinite. In any complete naming structure **any** subset of the domain will have a possible name corresponding to it. Of course, it is not the case that there is a possible world which contains **all** such names simultaneously, but that is not a problem for the modal interpretation of higher-order quantification. The initial plausibility of the objection results from the ambiguity between:

> (8.13) It is possible that for every subset of the domain there is a name which names all and only those objects which are elements of this subset.

and

> (8.14) For any subset of the domain it is possible that there is a name which names all and only all its elements.

Let us grant that the set of names in a possible world has to be countable; at least on the nominalist view names are finite inscriptions or utterances over a finite alphabet. The difference between the readings given above is pretty much like the difference between the statement that it is possible that every real number has a name (which would require the existence of uncountably many names) and the claim that for any real number it is possible for it to have a name. The first is false, the second true. **Size limitation objections** may work well against (8.13). But it is not (8.13) but rather (8.14) that is posited in the modal interpretation under discussion, and it is not prone to this sort of difficulties.

8.8.5 Why not Go with Boolos's Semantics?

Another question which can be asked is how the currently defended approach improves on a semantics for plural quantification developed by Boolos (1998a). Strictly speaking, Boolos was talking about plural quantification for the language of set theory in which individual variables ranged over sets and the question was whether second-order variables have to be taken as ranging over classes (proper classes included). But accommodating this semantics to the case where individual variables range over individuals and the question is whether second-order variables have to be taken as ranging over sets is rather straightforward.

In such a semantics, satisfaction of a formula is defined relative to a domain D,[32] a sequence s (an assignment), and what Boolos calls 'a new [...] second-order variable "R".' (Boolos 1998a, 336) The key clauses are:

$$\langle D, s, R \rangle \models Vv \text{ iff } R\langle V, s(v) \rangle$$

$$\langle D, s, R \rangle \models \exists v \, \phi \text{ iff for some object } x \in D, \text{ there is a sequence } t$$
$$\text{which agrees with } s \text{ on all variables different}$$
$$\text{from } v, \text{ such that } t(v) = x \text{ and } \langle D, t, R \rangle \models \phi.$$

$$\langle D, s, R \rangle \models \exists V \, \phi \text{ iff there is a T such that for any second-order}$$
$$\text{variable } U \text{ other than } V, \text{ for any } x \in D,$$
$$(T\langle U, x \rangle \equiv R\langle U, x \rangle), \text{ and} \langle D, s, T \rangle \models \phi.$$

Boolos comments on the status on the "new second-order variables" employed in the metatheory:

> The present theory, however, makes no explicit mention of sequences whose values are (proper) classes [in our modified case: sets – RU]. It does not proceed by introducing functions that assign to each second-order variable a unique class, possibly proper. Instead it employs a new predicate which, as one may say, is true or false relative to [...] a sequence of first order variables, and some (or perhaps no) ordered pairs of second-order variables and sets [in our case: individuals – RU] to the second-order variable R. There is, however, no need to take the theory as assigning classes [in our case: sets – RU] [...] to the second-order variables (Boolos 1998a, 336–337).

The essence of the proposal is to allow plural interpretations of second-order variables, on which a second-order variable takes individuals as its values, but no restriction is put on the number of its values.

While the nominalist will applaud the idea that second-order variables do not have to range over sets, understanding the Boolean semantic clause for second-order quantification nominalistically might be problematic. How is the nominalist to understand 'there is a T' and those 'new second-order variables' in general?

---

[32] Strictly speaking, Boolos takes the domain to be fixed, but relativizing to a domain is a rather trivial part of my straightforward modification of Boolos's semantics.

- One way to go is to take them to be binary predicates truly predicable of a second-order variable and an object. On this reading, '$R\langle v, x\rangle$' simply reads 'x is among the values of the variable v'. The problem is, if the quantification is taken to range over predicates, the size limitation objection strikes back: there can be at most countably many predicates (even worse, only finitely many predicates exist).

- Another way is to take them to plurally refer to ordered pairs of second-order variables and individuals. The problem is, in this case '$\exists V\,\phi$' reads 'there are some ordered pairs of variables and individuals T such that...', and it is rather unclear whether a nominalist who wants to dispose of classes will be fine with ordered tuples as referents.

- Perhaps, given that the nominalist should at least admit that certain predicates/names sometimes refer to certain things, one could read 'for some T' as 'for some reference relation'. But caution is advised—the nominalist usually doesn't think that reference relations float around without tokens. Thus, it might be philosophically more transparent to bring expression tokens into the picture (but then, due to things like the size limitation objection, going modal seems advisable). But any attempt to clear up such issues is very likely to lead into the general direction sketched in this paper.

- Last but not least, it is not clear how the Boolean semantics can be generalized to yet higher-order of quantification. In contrast, bringing tokens into the picture allows for further introduction of tokens referring to tokens, and so on (see (Urbaniak 2008c) and (Urbaniak 2010) for such a generalization of the framework developed in this chapter).

8.8.6 Why Complete Naming Structures?

When we try to provide a nominalist story about a logic equivalent to QNL with set-theoretic semantics, what reasons do we have to postulate that it is only complete naming structures that matter? In a sense, why do we concern ourselves with such a strong notion, one might say? If the claim is that this is what nameability really is like, one might launch an attack along the following lines:

> Certain objects are nameable if and only if there is a person who could name them. But it is far from obvious that for any objects there is a person who can name them. For suppose a person utters (or writes) a token and wants to establish its reference. It seems that she either has to be able to formulate in her language a description which only those objects satisfy and which employs only those expressions whose reference she may take for granted, or she has to have direct epistemic access to those objects (well, at least pretty much as direct as the act of pointing at an object, as opposed to determining reference through a description). But it is not obvious that her language is sufficiently expressive to provide her with a description that determines this set, and it is quite obvious that her relation to quite a few objects in the world will be far from direct epistemic access. Hence, it seems, we should not attach that sort of significance to complete naming structures.

Here is a rejoinder to this sort of criticism. The required notion of nameability does not require the existence of persons who can actually name those objects, nor does it obviously require that there be an ideal agent who would be able to name certain objects, had he existed in this possible world. We can consider objects belonging to a certain set nameable if we know that by introducing a name that denotes all and only its elements we would not run into troubles of logical nature, and that is all there is to it (see also the remarks about mathematical theology on p. 196).

An analogy might help. Suppose someone argues:

> Certain functions are computable if and only if there is a person who could compute them. But it is far from obvious that for any (say) Turing computable function there is a person who can compute its value for any input. Quite obviously, there are certain computable functions and certain inputs such that no person alive would be able to compute the values of those function for those inputs. Therefore we should not attach such significance to the class of all Turing-computable functions.

Arguments of this sort do not disprove the usefulness of the notion of Turing-computability. At best, they just show that people's actions are not exactly what Turing computability theory is about. Also, even though the notion of computability used here is indeed quite strong and therefore it may aid the imagination to think of it as a theory of an ideal computer (in the good old sense of the word, when 'computer' meant the same as 'reckoner' or 'figurer'), ideal computing agents are not exactly the subjects of the discipline.

Let me draw another parallel here. Of high importance in Turing computability theory are limitation theorems: the knowledge that **even if** we put pretty much no practical restrictions on computability there still are functions which cannot be computed is quite substantial. The purpose of the nameability theory is to provide an (in a sense) substitutional semantics for QNL. Note that when we think of a logic substitutionally one of our main concerns is to be able to apply it to any names whatsoever without any uncertainty as to whether the validity or truth will be preserved when we eliminate the universal quantifier and substitute a name for the variable that it binds. Here, widening the range of possible names contributes to the universality of the logic.

8.8.7 Set-Theoretic Content of Plural Logic

Yet another style of arguments against the belief that the logic of plurals (or QNL for that matter) is ontologically neutral can be extracted from an argument against the *logicality thesis* ("the logic of plurals (or QNL) qualifies as pure logic"). It is not my present concern to define (or discover) what being a pure logic consists in; let me just say that the main feature of plural logic that struck some people as problematic is its apparently set-theoretic content. The claim that QNL has a set-theoretic content, together with the initially plausible assumption that whatever has set-theoretic content is committed to sets lead to the conclusion that QNL is committed to sets.

Here is a good example of a discussion pertaining to the logicality of the logic of plurals:[33]

> First [in the logic of plurals] there are the tautologies and the inference rules governing identity and the singular quantifiers. There is broad consensus that these qualify as logical. Next there are the inference rules governing the plural quantifiers. Since these rules are completely analogous to the rules governing the singular quantifiers it would be hard to deny that they too qualify as logical. Then there are the extensionality axioms and the axiom that all pluralities are non-empty. These axioms are unproblematic because they can plausibly be taken to be analytic. What remains are the plural comprehension axioms, where things are much less clear. For these axioms have no obvious singular counterparts, and their syntactical form indicates that they make existential claims.[34] (Linnebo 2004).

Mutatis mutandis, this objection can be mounted against the ontological innocence of QNL: one might say, it not only makes existence claims (comprehension axioms of the form $\exists a \, \forall b \, (b \, \varepsilon \, a \equiv b \, \varepsilon \, a \wedge \phi(b))$ come out true in any complete naming structure), but also clearly has non-logical content because the comprehension axioms have no corresponding first-order (or singular) counterparts. Both qualms deserve consideration. Let us start with the first one.

If we believe that nameability is quite well-understood without postulating abstract objects and we accept the modal reading of QNL quantifiers, there is no reason to read any serious existential import into comprehension axioms. They just mirror the fact that for any formula ϕ of QNL at any possible world w there is no logical obstacle to introducing a new name that would refer to those individuals in w which satisfy ϕ the fact that names which coincide in reference with one-place open formulas of QNL are logically admissible.[35]

Worries about the apparent existential form of comprehension notwithstanding, there is yet another intuition behind this line of criticism. If something is a fairly obvious generalization of what is already present in the standard first-order logic, its nature is logical. Thus, for instance (at least when it comes to the rules of inference in any standard axiomatization) the rules pertaining to behavior of quantifiers in QNL are on par with the rules pertaining to quantifiers in first-order logic. The same goes for rules pertaining to Boolean connectives. Comprehension axioms, on the other hand, one might say, are something new: there is nothing in first-order logic which suggests anything like comprehension axioms in QNL.

This line of attack can be circumvented by describing a singular variant of comprehension. The essence of comprehension, as I take it, is that forming new definitions of a certain sort does not lead outside of the quantifier scope. Now, suppose we take the first-order language, but instead of giving it the standard set-theoretic semantics, we develop for it a modal-substitutional semantics of the sort developed in this paper.

---

[33] Linnebo (2003) considers also another argument against the ontological innocence of the logic of plurals. Since it rather pertains to higher-order extensions of the logic of plurals, I will not talk about it.

[34] Linnebo remains neutral when presenting this argument, I just use his wording for its brevity and clarity

[35] Say we take $\phi(a)$ to refer to those objects whose singular names (had there been any) substituted for a in $\phi(a)$ (which does not contain other free variables) would yield true sentences.

In such a context, the first-order quantifiers range over possible ways singular terms can be. Within this setting, it is rather clear what singular comprehension would do: it would require that for any possible world, for any individual which can be described in that language (that is, for any formula, which is satisfied by exactly one object), it is possible to introduce a singular term referring to that individual.

8.8.8 Is Multiple Reference Relation Innocent?

Granted that quantification over sets has been eliminated in virtue of the modal-substitutional interpretation of plural quantification, one can still raise doubts regarding the innocence of the reference relation that possible names bear to objects in the world. One could argue along the following lines:[36]

> Ok, so we made names the values of variables. How does it help if, it seems, the names themselves can still refer to sets *via* the δ relation? Why isn't this just syntactic sugar?

To understand the misreading that stands behind this sort of objection let us first note that there are three, not two, readings involved. One, the straightforwardly set-theoretic one reads '$\exists a$' as:

> (8.15) There is a set a such that...

The second one, which is the reading implicit in the above objection interprets '$\exists a$' as:

> (8.16) It is possible to introduce a name which refers to a certain set such that...

Of course, this second reading of quantifiers, even though, in a sense, substitutional, does not eliminate ontological commitment to sets. The natural language reading that the modal semantics I developed suggests is none of the above. Rather, it interprets the phrase in question as:

> (8.17) It is possible to introduce a name (which may not refer at all, refer to a unique object or refer to multiple objects), such that ...

The last reading does **not** interpret possible names as referring to sets. Rather, it assumes that the fact that a name (a countable noun phrase) refers to multiple objects or does not refer to anything at all is of the same ontologically innocent status as the fact that a name refers to a unique object.

Even the most severe opponents of the ontological innocence of the logic of plurals allow that reference of singular terms does not bring in any commitment to sets. So, it seems, the reference relation alone is not a problem for a nominalist. Now, why should this change when he admits that a name can fail to refer to anything? Well, it should not. To say that 'Spiderman' does not refer to anything is **not** to say that 'Spiderman' refers to the empty set. It is only a very prejudiced commitment to the

---

[36] Richard Zach turned my attention to this sort of doubt.

'Fido'-Fido principle that may convince one that for any expression there has to exist its unique referent which explains how the expression contributes to the truth-value of sentences in which it occurs. Having said this, let us ask: Why should the situation be different when we turn to names that refer to more than one object?

It seems that competent language users which did not have any experience with philosophy should have a say in this debate; and I presume that very few of them would suggest that 'Dogs are animals' is true in virtue of there being certain objects which do not exist in space and time[37] such that each dog remains in a certain relation of "belonging" to one of them and each animal remains in this relation to another and those two untouchable objects remain in a certain unobservable relation. To the contrary, the fact that underlies the truth of this sentence is quite observable and pertains to certain real individual objects: dogs and animals. It requires a certain amount of philosophical sophistry to deny that 'dogs' refers to dogs or that 'animals' refer to animals along the lines suggested above.

When we try to explain nominalistically the logic of plurals, there are two aspects that have to be dealt with. First, we should say what plural quantifiers quantify over (or how they should be interpreted). My reading is: they are modal devices which are explained substitutionally in terms of possible countable noun phrases. Second, the very use of plurals, even without quantifiers should be explained as nominalistically acceptable. My explanation is: unless someone assumes that every expression should have a unique referent and believes in set theory already there is no linguistic evidence to the effect that the existence of sets is required to make sense of the reference of plurals. Interpreting plurals as referring to sets has certain theoretic advantages when we do formal semantics (set-theory is fairly well-known and the semantics is slightly simpler), but this does not constitute any argument to the effect that this is the only way to make sense of plural constructions. A more parsimonious answer is: plurals refer to objects directly. 'Dogs' refers to dogs, 'chairs' refers to chairs and it is enough to believe in the existence of chairs and dogs (as opposed to believing in the existence of sets as well) in order to understand sentences in which 'dogs' or 'chairs' occur.

The expression 'refers to' in the relevant sense has pretty much the same meaning as 'is truly predicable of'. When we introduce a name for already existing objects, it refers to those already existing objects of which it is truly predicable. 'Rafal Urbaniak' refers to me because 'This is Rafal Urbaniak' is true when the utterer points at me. There is nothing that forces us to say that being-truly-predicable-of in the case of 'Rafal Urbaniak' requires that there be a singleton set containing me as its sole element. What is more, clearly if being-truly-predicable-of something is a necessary condition for referring to this thing (and this is how I prefer to understand the reference relation), 'Rafal Urbaniak' does not refer to any set (even though being an eternal being might be an upside of being a set, I tend to think of myself as an object which is not out of space and time). Similarly, if we have an empty name, like 'Superman', there is no object of which we can truly say that it is Superman ('Superman' is not

---

[37] For the sake of simplicity I ignore the position that sets exist in space and time in addition to usual objects as quite unbelievable. This position was held for a while by Maddy (1980). A plausible criticism of it is to be found in Chihara (1990, 194–215).

truly predicable of the empty set). It is false that in this sense 'Superman' refers to the empty set because it is false that the empty set is Superman. How about names that intuitively name more than one object? The expression 'chair(s)' is truly predicable of each and every chair and therefore every chair is a referent of the word 'chair'.[38] Also, the set of chairs (if there is such an object) is not a referent of the word 'chair(s)', for it is not a chair itself. Hence, in this sense of the phrase, reference of names is a relation between tokens and individuals themselves which does not proceed *via* sets and which does not require the existence of sets.

8.8.9 Counting Reference Relations

A slightly related worry has been raised by Hochberg (1984). In the beginning of his paper, he declares:

> One attempt to avoid a platonistic ontology involves using the notion 'is true of'. Using this notion, one claims that a predicate, say 'white' is true of a particular object whereas a sign such as a proper name refers to the object. This is done to avoid holding (1) that a predicate refers to a universal property while the proper name refers to an individual and (2) that a sentence used to predicate the one sign of the other indicates the fact that the individual exemplifies the property. Thus, an attempt is made to offer a coherent ontology recognizing only individuals. If cogent, this gambit provides a more parsimonious ontology than platonistic alternatives. Here I shall argue that the gambit fails (Hochberg 1984, 150).

The core of the objection is that when the Platonists gives an account of the truth-conditions of sentences like 'a is W' (where 'a' is a singular term, and 'W' a predicate), they're using one reference relation (the constant refers to an object, the predicate refers to a property, and the sentence is true iff a possesses the property W), whereas the nominalist requires two different reference relations: one for the singular terms (reference), and one for the predicates (being true of), and thus "the nominalistic gambit is no more parsimonious than the platonist's" (p. 151).

The main reason why Hochberg thinks that different relations are involved is that claims involving the Platonist's reference relation, like:

'a' refers to a
'W' refers to W

display "an obvious triviality or emptiness", reflecting "the redundancy that a label labels what it labels. They can, with appropriate rules, be seen to be true by inspection" (p. 151), whereas:

'W' is true of a

"…is quite different. It holds when the individual has the property, and hence does not merely reflect our having interpreted certain signs" (*ibidem*).

---

[38] Thus, on the view under discussion, the reference of distributive plurals is nothing above and beyond the reference to particulars. This means that we also do not need anything like "mereological fusion of all chairs" as the referent of 'chair(s)' (at least as far as plural quantification is concerned).

There are a few moves the nominalist can make in response to this sort of criticism. First of all, they might point out that the weight of someone's ontological commitment is not to be measured by counting the relations: as long as the relations are nominalistically acceptable, the nominalist can admit as many of them as he pleases without running into any immediate difficulty just because of the number of relations that he admitted.

Secondly, they might point out that a rather uncharitable principle of counting relations has been used in the argument: it doesn't seem to hold in general that if in certain cases it is more difficult do ascertain whether a certain relation holds, there really are two relations: one of the easier cases, and one for the more difficult ones.

Thirdly, they might challenge the claim that there really is a deep epistemological difference between the cases that Hochberg finds so radically different. Contra Hochberg, claims like:

'a' refers to b

where 'a' is a singular term and 'b' is a singular term referring to the same object, are far from being trivial. Actually, their lack of obviousness is a source of many philosophical puzzles pertaining to naming, propositions and propositional attitudes.

Perhaps, the difference that Hochberg had in mind is that singular terms refer directly, whereas multiply referring terms seem to refer *via* properties, and since the nominalist doesn't admit properties, they have to at least admit that the reference relation is different. This doesn't sound exactly right: the distinction between having and not having a descriptive content cuts across the distinction between singular and plural terms. I can easily point at the three chairs standing in my room and christen them directly with a non-descriptive plural term. Anyway, even if there is a difference, the nominalist can still express this fact that it is not two distinct reference relations, but rather one reference relation holding between different some objects for more complicated reasons.

Of course, the nominalist should tell a story about how to interpret predication of descriptive terms without invoking properties (this lies beyond the scope of my current considerations), but this is a different challenge from claiming that multiple relation itself is a problem.

8.9 Loose Ends

Let me just end with briefly reminding you what the main open tasks for the modal nominalist are:

- The framework needs to be extended to the full-blown language of Ontology. Arguably, once the framework of modal semantics is allowed, this is rather straightforward.
- The semantic account, if possible, should be superseded by an axiomatization independent of set theory, given together with a philosophical defense of the

nominalistic acceptability of the axioms involved. This would definitely eradicate concerns related to the role of set theory in the account. It is not known whether this can be done.

- It would also be nice if the axiomatization could avoid the possible-world discourse, so that no worries about commitment to the existence of possible worlds arise. Again, whether this can be achieved is not known. (See however Chihara (1998) for an attempt.)
- All this would result only in a theory capable of emulating set theory. Yet, there are standard concerns related to whether other mathematical objects can be plausibly identified with sets (Benacerraf 1965), because there are often too many ways this can be done for any of them to count as a compelling philosophical story. Thus, a sensible way of extending the nominalistic framework to mathematical theories other than set theory (other than emulation within set theory) should be provided.

While I do think these desiderata can be met, it is a story for a completely different book.

References

Austin, J. L. (1966). Three ways of spilling ink. *Philosophical Review*, *75*(4), 427–440.

Benacerraf, P. (1965). What numbers could not be. *Philosophical Review*, *74*, 47–73.

Benacerraf, P. (1973). Mathematical truth. *Journal of Philosophy*, *70*, 661–679.

Boolos, G., Burgess, J. P., & Jeffrey, R. C. (2002). *Computability and Logic*, 4th edn. Cambridge: Cambridge University Press.

Boolos, G. (1998b). On second-order logic. In R. Jeffrey (Ed.), *Logic, logic, and logic* (pp. 37–53). Cambridge: Harvard University Press.

Boolos, G. (1998a). Nominalist platonism. In R. Jeffrey (Ed.), *Logic, logic, and logic* (pp. 73–87). Harvard: Harvard University Press.

Burgess, J. P. (2004). E pluribus unum: Plural logic and set theory. *Philosophia Mathematica*, *12*, 193–221.

Chihara, C. S. (1973). *Ontology and the vicious-circle principle*. Ithaca, NY: Cornell University Press.

Chihara, C. S. (1990). *Constructibility and mathematical existence*. Oxford: Oxford University Press.

Chihara, C. S. (1998). *The worlds of possibility: Modal realism and the semantics of modal logic*. Oxford: Oxford University Press.

Dunn, J. M., & Belnap, N. D. (1968). The substitution interpretation of the quantifiers. *Nous*, *2*(2), 177–185.

Hellman, G. (1989). *Mathematics without Numbers: Towards a modal-structural interpretation*. Oxford: Oxford University Press.

Hochberg, H. (1984). Nominalism, platonism and being true of. *In logic, ontology and language. Essays on truth and reality*. Munchen: Philosophia Verlag.

Kearns, J. (1962). *Lesniewski, language, and logic*. PhD thesis, Yale University, New Haven.

Kearns, J. (1969). Two views of variables. *Notre Dame Journal of Formal Logic*, *10*, 163–180.

Kielkopf, C. (1977). Quantifiers in ontology. *Studia Logica*, *36*, 301–307.

Kripke, S. (1976). Is there a problem about substitutional quantification? *truth and meaning; essays in semantics* (pp. 324–419). Oxford: Clarendon Press.

Küng, G., & Canty, J. T. (1970). Substitutional quantification and Leśniewskian quantifiers. *Theoria*, *36*, 165–182.

Küng, G. (1974). Prologue-functors. *Journal of Philosophical Logic*, *3*, 241–254.

Küng, G. (1977). The meaning of the quantifiers in the logic of Leśniewski. *Studia Logica*, *36*, 309–322.

Lejewski, C. (1954b). Logic and existence. *The British Journal for the Philosophy of Science*, *5*, 104–119.

Leśniewski, S. (1927). O Podstawach Matematyki, Wstęp. Rozdział I: o pewnych kwestjach, dotyczcl II: o 'antynomji' p. Russella, dotyczacej 'klasy klas, nie będacych własnemi elementami'. Rozdział III: o różnych sposobach rozumienia wyrazów 'klasa' i 'zbiór'. *Przeglad Filozoficzny*, *30*, 164–206 [On the foundations of mathematics. Introduction. Ch. I. On some questions regarding the sense of the 'logistic' theses. Ch. II. On Russel's 'antinomy' concerning 'the class of classes which are not elements of themselves'. Ch. III. On various ways of understanding the expression 'class' and 'collection' (Lesniewski, 1991, 174–226)].

Linnebo, Ø. (2004). Plural quantification. In E. N. Zalta (Ed.), *The stanford encyclopedia of philosophy*. The Metaphysics Research Lab, Stanford: Stanford University.

Linnebo, Ø. (2003). Plural quantification exposed. *Nous*, *37*, 71–92.

Maddy, P. (1980). Perception and mathematical intuition. *Philosophical Review*, *89*, 163–196.

Marcus, R. B. (1972). Quantification and ontology. *Nous*, *6*, 240–250.

Parsons, C. (2008). *Mathematical thought and its objects*. Oxford: Oxford University Press.

Prior, A. (1965). Existence in Leśniewski and Russell. In J. Crossley (Ed.), *Formal systems and recursive functions* (pp. 149–155). Amsterdam: North-Holland.

Quine, W. V. (1947). On universals. *The Journal of Symbolic Logic*, *12*, 74–84.

Quine, W. (1952). Review: On the notion of existence. Some remarks connected with the problem of idealism, by Kazimierz Ajdukiewicz. *The Journal of Symbolic Logic*, *17*, 141–142.

Quine, W. (1970). *Philosophy of logic*. Englewwod Cliffs: Prentice-Hall.

Rickey, F. (1985). Interpretations of Leśniewski's ontology. *Dialectica*, *39*(3), 181–192.

Sagal, P. (1973). On how best to make sense of Leśniewski's ontology. *Notre Dame Journal of Formal Logic*, *14*(2), 259–262.

Shapiro, S. (1993). Modality and ontology. *Mind*, *102*, 455–481.

Simons, P. (1985). *A semantics for ontology. Dialectica*, *39*(3), 193–215.

Simons, P. (1995). Lesniewski and ontological commitment. In D. Miéville & D. Vernant (Eds.), *Stanislaw Lesniewski Aujourd'hui, number 16 in recherches philosophie, langages et cognition* (pp. 103–119). Grenoble: Université de Grenoble.

Smith, P. (2009). Charles Parsons: Mathematical thought and its object. http://www.phil.cam.ac.uk/teachingstaff/Smith/logicmatters/Resources/Parsons1.pdf

Tomberlin, J. E. (1997). Quantification: objectual or substitutional? *Philosophical Issues*, *8*, 155–167.

Urbaniak, R. (2008). Reducing sets to modalities. In Reduction and elimination. Proceedings of the 31st Wittgenstein, Symposium, pp. 359–361.

Urbaniak, R. (2010). Neologicist nominalism. *Studia Logica*, *96*, 151–175.

Woleński, J. (1992). Review: Constructibility and mathematical existence by Ch Chihara. *History and Philosophy of Logic*, *13*, 233–234.

Index

R. Urbaniak, *Leśniewski's Systems of Logic and Foundations of Mathematics*,
Trends in Logic 37, DOI: 10.1007/978-3-319-00482-2,
© Springer International Publishing Switzerland 2014